Library of
Davidson College

INTENSIONAL AND HIGHER-ORDER MODAL LOGIC

In memory of
Richard Montague

NORTH-HOLLAND
MATHEMATICS STUDIES   19

Intensional and Higher-Order Modal Logic
With Applications to Montague Semantics

DANIEL GALLIN
Department of Mathematics
University of San Francisco
San Francisco, California, USA

1975

NORTH-HOLLAND PUBLISHING COMPANY - AMSTERDAM • OXFORD
AMERICAN ELSEVIER PUBLISHING COMPANY, INC. - NEW YORK

© NORTH-HOLLAND PUBLISHING COMPANY – AMSTERDAM, 1975

All Rights Reserved. No part of this publication may be reproduced, stored in a retrieval system, or transmitted, in any form or by any means, electronic, mechanical, photocopying, recording or otherwise, without the prior permission of the copyright owner.

North-Holland ISBN: 0 7204 0360 X
American Elsevier ISBN: 0 444 11002 X

Published by:
NORTH-HOLLAND PUBLISHING COMPANY – AMSTERDAM
NORTH-HOLLAND PUBLISHING COMPANY, LTD. – OXFORD

Distributors for the U.S.A. and Canada:
American Elsevier Publishing Company, Inc.
52 Vanderbilt Avenue
New York, N.Y. 10017

PRINTED IN THE NETHERLANDS

PREFACE

In a series of papers written during the period 1967-1971, Richard Montague outlined a highly original approach to the problem of providing a precise account of natural language syntax and semantics. In a sharp departure from the linguistic methods of the Chomsky school, Montague introduced a powerful body of techniques from the field of mathematical logic, principally the set-theoretic semantical methods pioneered by his teacher Tarski.

Montague's tragic death in 1971 cut short what was certainly the most ambitious research undertaking of his career, and one for which he was uniquely qualified. Although he completed only three papers dealing specifically with natural language, the ideas they contain have provided the basis for an entire branch of current linguistic research, and the interest in his work continues to grow among philosophers, linguists and logicians. The present work attempts to provide the technical background necessary for a thorough understanding of Montague Semantics, at the same time exploring some of the mathematically interesting applications of higher-order modal logic.

The focus of Part I is the logic of intensions, denoted by IL, which Montague introduced in his paper "Universal Grammar." This system extends Church's functional theory of types by the addition of two operators, corresponding roughly to intension and extension. Montague's formalized English fragments admit translation into IL, which is given a "possible worlds" semantics along the lines of Carnap-Kripke.

Following a brief introduction to the Montague program in Chapter 1, the syntax and semantics of IL are set out in detail. A natural axiomatization is provided, and Henkin's generalized completeness theorem for the theory of types is extended to the Montague system. This leads to a standard completeness theorem for a restricted class of "persistent" formulas, a result which has applications to certain "extensional" fragments of English.

In Chapter 2 some natural axiomatic extensions of IL are considered and normal forms are obtained for formulas of IL. In addition, Montague's system is compared with a two-sorted extensional theory of types.

Part II, which is essentially self-contained, deals with an alternative formulation of higher-order modal logic, denoted by ML_p. This system takes quantifiers and the necessity operator as primitives and allows only predicate types, in distinction to the arbitrary functional types of IL. Although equivalent to Montague's system, ML_p is perhaps more natural to the logician, and it has a number of interesting applications of its own in modal logic and set theory. Bressan has shown that such systems are also of interest in connection with the foundations of physics.

In Chapter 3, generalized completeness is proved for ML_p and for the theory ML_p+C obtained by adding a natural axiom schema of comprehension. A related principle of extensional comprehension, first proposed by Bressan, is shown to be equivalent in ML_p+C to an axiom of atomic propositions considered by Kaplan and Fine. Every general model of ML_p is shown in §10 to be homomorphic, in a truth-preserving sense, to one in which any two indices (possible worlds) are distinguishable by a formula. In §12 a general theory of propositional operators is developed within ML_p which includes "axiomatically" defined classes of operators and those arising from Kripke-type relevance relations as special cases.

In Chapter 4 a Boolean semantics is defined which validates every theorem of ML_p+C. This semantics is applied to show the independence of the extensional comprehension principle from the axioms of ML_p+C, and to obtain a number of other independence results in higher-order modal logic. Topological models, in the sense of McKinsey and Tarski, are explored in §16, and in §17 the Boolean semantics for ML_p is combined with the earlier generalized semantics to reconstruct the Scott-Solovay proof of Cohen's result on the independence of the continuum hypothesis. In this application of higher-order modal logic to set theory, certain modal sentences function as "interpolants" which express in formal terms various properties of the underlying Boolean algebra.

Except for minor revisions, the present work constituted my doctoral dissertation in mathematics, submitted to the University of California, Berkeley, in September 1972. I began working with Professor Montague

PREFACE

in July 1970, investigating several questions related to his system IL. Our work was interrupted by his death in March 1971, and Professor Dana Scott generously agreed to supervise the completion of my dissertation, for which I am deeply appreciative. I am also greatly indebted to the other members of my doctoral committee, Professors Leon Henkin and Robert Vaught, for their consistent direction and advice.

I must thank in addition Nuel Belnap, Harry Deutsch, Haim Gaifman, David Kaplan, Uwe Mönnich, Barbara Partee and Robert Solovay for helpful conversations and correspondence, my wife Janet for her patience, and the National Science Foundation for providing financial support during 1970-1971 under N.S.F. Science Faculty Fellowship No. 60068.

Montague's semantical methods are coming to seem less formidable, thanks largely to the efforts of Barbara Partee and others to bridge the separate disciplines of linguistics, philosophy of language, and mathematical logic. One is encouraged to hope that the work of Richard Montague may eventually bring these disciplines closer to their common goal, the understanding of language.

Daniel Gallin

University of San Francisco
June 1975

CONTENTS

PART I.   INTENSIONAL LOGIC

CHAPTER 1.   INTENSIONAL LOGIC

§1.   Natural Language and Intensional Logic 3
§2.   The Logic IL 10
§3.   Generalized Completeness of IL 17
§4.   Persistence in IL 37

CHAPTER 2.   ALTERNATIVE FORMULATIONS OF IL

§5.   Modal T-Logic 41
§6.   Extensions of IL and ML_T 44
§7.   Normal Forms 53
§8.   Two-Sorted Type Theory 58

PART II.   HIGHER-ORDER MODAL LOGIC

CHAPTER 3.   HIGHER-ORDER MODAL LOGIC

§9.   Modal Predicate Logic 67
§10. Propositions in ML_p 79
§11. Atomic Propositions and EC 84
§12. Propositional Operators 89
§13. Relative Strength of IL and ML_p 98

CHAPTER 4.   ALGEBRAIC SEMANTICS

§14. Boolean Models of ML_p 106
§15. Modal Independence Results 112
§16. Topological Models of ML_p 122
§17. Cohen's Independence Results 132

Bibliography 144

PART I. INTENSIONAL LOGIC

CHAPTER 1.   INTENSIONAL LOGIC

§1. Natural Language and Intensional Logic

When we speak of a theory of meaning for a natural language such as English, we have in mind an analysis which obeys the functionality principle of Frege, according to which the meaning of a given expression should be a function of the meanings of its constituents.[1] Philosophers of language since Frege have accepted the distinction, in discussions of meaning, between the <u>extension</u> or denotation of an expression ζ , and its <u>intension</u> or sense. Let us denote the former by $\text{Ext}[\zeta]$, the latter by $\text{Int}[\zeta]$. We know what the extensions of certain sorts of English expressions should be, according to semantical conventions which we tacitly accept when we translate English sentences into the symbolism of predicate logic. For example, if ζ is a name (e.g., 'Jones') and we denote the universe of all individuals by D , then $\text{Ext}[\zeta]$ is an individual, i.e., an element of D . If ζ is a common noun phrase (e.g., 'former thief') or an intransitive verb phrase (e.g., 'run slowly'), then $\text{Ext}[\zeta]$ is a set of individuals, or equivalently, an element of the set 2^D (the set of former thieves, the set of individuals who run slowly).[2] If ζ is a sentence then $\text{Ext}[\zeta]$ is simply a truth-value, or element of the set $2 = \{0,1\}$.

Can we identify meaning with extension? For the purposes of a beginning course in symbolic logic, we often do. Thus, for example, the extension (truth-value) of the sentence 'Jones runs slowly' depends only on the extensions of 'Jones' and 'run slowly', construed as above, and Frege's

---

[1] Frege [1892], English transl. in Feigl and Sellars [1949]. Carnap [1947] refers to "Frege's principles of interchangeability."

[2] Here and throughout, exponentiation of sets has its usual meaning: B^A denotes the set of all functions, or mappings, from A into B. As usual, we identify the number 2 with the set $\{0,1\}$.

functionality principle is satisfied. Philosophers have long been aware, however, of natural examples for which Frege's principle fails when meaning and extension are identified, examples involving "oblique" or intensional contexts.[3] For example, in the sentence 'Necessarily the morning star is identical with the morning star', which should certainly be counted true if we understand "necessarily" to mean "in all possible worlds," replacement of the second occurrence of the constituent 'the morning star' by the expression 'the evening star', which has the same extension, produces the false sentence 'Necessarily the morning star is identical with the evening star' -- false because we can easily imagine a world in which these stars are not identical. Thus the extension of the sentence is not a function of the extensions of its parts. To take another example, let us suppose that Jones is at this moment a member of the United States Senate, so that the common noun phrase 'colleague of Jones' has the same extension as the phrase 'United States Senator'. The compound phrase 'former colleague of Jones' has as its extension a certain set of individuals, among whom is Jones's old law partner, Smith. If we replace the constituent 'colleague of Jones' by the coextensional 'U.S. Senator', however, we obtain the phrase 'former U.S. Senator', the extension of which does not contain Smith. This shows that the extension of 'former colleague of Jones' is not a function merely of the extensions of its parts, so that Frege's principle fails here as well. This second example shows that the difficulty cannot be avoided by simply refusing to countenance the existence of such entities as "possible worlds," for even in the absence of such notions we encounter problems of the same sort.

Aware of the special problems posed by oblique contexts, Frege did not abandon the functionality principle for extensions but rather held the view that the extension of an expression ζ depends on the syntactic context in which it occurs. When ζ is used in an ordinary context its extension is $\text{Ext}[\zeta]$, but when used in an oblique context its extension becomes $\text{Int}[\zeta]$, the (ordinary) sense of ζ. According to this view, the failure of the functionality principle for extensions is attributable to the ambiguity of natural language. In principle one could eliminate the difficulties by introducing, for each expression ζ, a new expression

---

[3] See Quine [1960] for discussion and examples.

$^\wedge\zeta$ (the <u>concept of</u> ζ) whose extension is the intension of ζ . The functionality principle for extensions could then be preserved by using $^\wedge\zeta$ in place of ζ when ζ occurs in oblique contexts. Equivalently, one could amend the functionality principle for extensions to assert that the extension of a compound expression is a function of the extensions of those of its constituents standing within ordinary contexts, together with the intensions of those constituents standing within oblique contexts.

The first serious attempt to develop an <u>intensional logic</u> -- i.e., a logic of intensions -- along the lines suggested by Frege was that of Church [1951], who provided an axiomatic version of the Frege theory. The task of providing a referential semantics[4] for the theory remained, however, due to the problem of finding an adequate interpretation for intensions. In recent years this problem has been solved by means of a device which goes back to Carnap.[5]

Consider the examples given earlier, in which the functionality principle for extensions failed to hold. What seems essential in both examples is that the extensions of the various expressions involved depend on the particular <u>state of affairs</u>. For instance, the extension of the name 'the morning star' depends on the particular world in question; the extensions of the common noun phrases 'colleague of Jones' and 'U.S. Senator' both depend on the particular moment in time. When we know the extension of such an expression for a particular state of affairs we may still need to know its extensions for other states of affairs before we can determine the extension of a phrase of which it is a constituent. For example, to determine the set of former colleagues of Jones (at this moment), it is not enough to know the set of colleagues of Jones (at this moment); we must also know the set of colleagues of Jones at each past moment. Carnap made the proposal that the intension of an expression be identified with the <u>function</u> on possible states of affairs whose value, at a particular state of affairs, is the extension of the expression in that state of affairs. That is, according to Carnap "extension" is really a function of two variables: we should speak of $\text{Ext}[\zeta,i]$, where ζ is an expres-

---

[4] I.e., a model theory in the sense of Tarski [1954].
[5] See Carnap [1947], §40.

sion and i is a possible state of affairs, in place of our earlier Ext[ζ] . We then identify Int[ζ] with the function F on the set I of possible states of affairs, such that F(i) = Ext[ζ,i] for each i ∈ I. This definition meets an implicit criterion of Frege's that the extension of an expression should be recapturable from its intension; in the present case the former is just the value of the latter at a particular state of affairs.[6]

Kaplan [1964] used Carnap's proposal to provide a semantics for Church's intensional logic. Kaplan followed a suggestion of Carnap, however, in identifying possible states of affairs with models of the underlying language, taking I to be a set of such models. This approach has a number of disadvantages, and it has now been widely supplanted by the approach of Kripke [1959], which takes the notion of a possible state of affairs -- in the context of modal logic, a possible world -- as primitive.[7] Montague [1968] and Scott [1970] suggested that a possible state of affairs be thought of as specifying the context of use appropriate to a particular language. Expressions of the language are thought of as indexical,[8] i.e., their extensions depend on a particular context of use. For example, to evaluate the truth or falsity of an utterance of the English sentence 'I was here yesterday', one must know the speaker s , the time t , the world w , and the spatial position p = (x,y,z) . Thus, a possible state of affairs -- in Scott's terminology an index or point of reference -- can be thought of as a sequence i = (s,t,w,p, ...) , where the remaining coordinates specify, in general, various other aspects of the context of use.[9] Given an expression ζ and an index i , we should be able to define the extension Ext[ζ,i] of ζ at i , and then, following Carnap, define the intension Int[ζ] of ζ by the equation Int[ζ](i) = Ext[ζ,i] for i ∈ I , where I is the set of indices.

What sorts of entities serve as the intensions of the particular types of English expressions considered earlier? Suppose we let D now

---

[6] See Church [1956], p. 9.
[7] Cf. Bayart [1958] for an early approach along these lines.
[8] The term is due to C.S. Peirce. See Bar-Hillel [1954].
[9] See the discussion in Lewis [1970].

represent the set of all possible individuals, i.e., individuals which exist with respect to at least one index $i \in I$. If ζ is a name then Int[ζ] will be an <u>individual concept</u>, i.e., an element of D^I. If ζ is a common noun phrase or intransitive verb phrase, Int[ζ] will be a <u>property of individuals</u>, i.e., an element of $[2^D]^I$. If ζ is a sentence, Int[ζ] will be a <u>proposition</u>, i.e., an element of 2^I.[10]

The potential which Carnap's idea holds for the analysis of natural language only becomes fully clear when we carry the construction to higher orders, for we can then assign intensions to many parts of speech whose analysis eludes ordinary predicate logic. Montague[11] first suggested such an approach for verbs which take a propositional object, such as 'know' and 'believe', which play a key role in a number of problems in the philosophy of language. Montague's proposal was to assign to such verbs intensions in the set

$$\left[2^{D \times 2^I} \right]^I ,$$

or in other words extensions which are functions mapping an individual and a proposition to a truth-value. The intension of the sentence 'Jones believes that snow is white' could be expressed in terms of the intensions of its constituents by the condition:

Int['Jones believes that snow is white'](i) =
    Int['believe'](i)(Int['Jones'](i),Int['Snow is white']) .

Although discussion of the proper treatment of belief continues among philosophers and seems to have no single satisfactory answer, Montague's higher-order intensional approach at least provides an adequate framework for the analysis of one sense of belief, according to which the object of belief is a proposition (as opposed to a sentence or some other entity).[12]

---

[10] Some authors, e.g., Scott [1970] and Lewis [1970], take concepts to be partial functions on I, possibly undefined for some indices.
[11] Talk given in 1967, reported in Montague [1970a].
[12] See Partee [1973].

Montague's student Kamp[13] pointed out that the intensions of adjectives could be taken to be elements of the set

$$\left([2^D]^{[2^D]^I} \right)^I ,$$

so that the extensions of an adjective map properties of individuals to sets of individuals. For the common noun phrase 'former colleague of Jones', for example, we would have the semantic condition:

Int['former colleague of Jones'](i) =
                    Int['former'](i)(Int['colleague of Jones']) .

In Montague [1970b] the method was extended to encompass a limited but completely formalized fragment of English, for which a precise syntax was provided in terms of grammatical categories and formation rules, as well as a referential semantics. The richer fragment of Montague [1970c] accommodates sentences, common and proper nouns, relative clauses, singular terms, adjectives, transitive and intransitive verbs (including so-called "intensional" verbs), and verbs taking a propositional object. The same paper also outlines a general theory of grammar and semantics.[14] In Montague [1973], the treatment is simplified somewhat and extended in various ways, e.g., to accommodate intensional prepositions and adverbs.

Various philosophers and linguists in recent years, among them Davidson, Parsons and Lewis, have been interested in giving a precise semantical account of natural language. The English fragments constructed by Montague constitute an important step in this direction. They are rich enough to provide an analysis of certain philosophical problems which hinge on intensional notions, and their underlying method has already been extended in various ways.[15] It is worth noting that the English fragments obtained in this way, as well as the formal logic IL to be discussed in §2, satisfy Frege's functionality principle for intensions -- the intension of a compound is a function of the intensions of its constituents -- and also the functionality principle for extensions, as amended to take into account

---

[13] Also, independently, Parsons [1968].
[14] Cf. Lewis [1970] and Partee [1975a].
[15] See Partee [1975b].

oblique contexts. Whether or not the intension Int[ζ] should be identified with the meaning of ζ is an arguable question; as Lewis points out, however, "intensions are part of the way to meanings ... and they are of interest in their own right." [16]

Let us use the term English to refer to one of the formalized fragments in Montague [1970c], [1973]. In each case we can single out a certain subset of the basic vocabulary which consists of extensional words: loosely speaking, those which do not create oblique contexts.[17] Consider, for example, the adjectives 'former' and 'tall'. Each has an intension in the set

$$\left([2^D] \; [2^D]^I \right)^I ,$$

but they differ in that the extension, at i, of 'former colleague of Jones' depends on the entire intension of 'colleague of Jones', whereas the extension, at i, of 'tall colleague of Jones' depends only on the extension at i of 'colleague of Jones'. Put differently, the set of tall colleagues of Jones, at i, is determined once we know the set of colleagues of Jones at i; the same is not true of the set of former colleagues of Jones at i. The adjective 'tall' is therefore extensional, in contrast to the intensional adjective 'former'. Extensionality of an adjective can be expressed by the following condition on its intension F:

If $i \in I$, $G, H \in [2^D]^I$, and $G(i) = H(i)$,

then $F(i)(G) = F(i)(H)$.

Similarly one can distinguish between extensional and intensional verbs, prepositions, etc., and require of every model of English that the intensions of all extensional words satisfy the appropriate semantic restriction. Those expressions which only involve extensional words[18] then comprise a sublanguage which we shall call Extensional English. We refer to this language briefly in §4.

---

[16] Lewis [1970], p. 25.
[17] See Montague [1970c], pp. 395-396.
[18] Certain of the formation rules may have to be restricted also.

§2. The Logic IL

Montague makes use, in Montague [1970c], [1973], of an auxiliary formal logic which he calls <u>Intensional Logic</u>, here denoted by IL. The language of IL is broad enough to encompass a wide variety of intensional notions; Montague uses the semantics for IL, in fact, to provide his semantics for English, by means of a translation of English into the symbolism of IL.[1] Our principal object of study in this chapter will be the logic IL.

In constructing domains of entities to serve as intensions of various expressions of English, we have observed the need to pass from a given domain A to another domain A^I, and from given domains A and B to the domain B^A. This observation motivates the choice of symbolism for IL, which is based on the theory of types as formulated in Church [1940], in which functional abstraction is taken as a primitive notion.

<u>Types</u>. Let e, t, s be any three objects, none of which is an ordered pair. The set of <u>types of</u> IL is the smallest set T satisfying:

(i) $e, t \in T$,

(ii) $\alpha, \beta \in T$ imply $(\alpha,\beta) \in T$,

(iii) $\alpha \in T$ implies $(s,\alpha) \in T$.

As will emerge later, objects of type e will be possible entities or individuals, objects of type t will be truth-values, objects of type (α,β) will be functions from objects of type α to objects of type β, and objects of type (s,α) will be functions from indices to objects of type α, i.e., senses appropriate to denotations of type α. We frequently write $\alpha\beta$ for (α,β) and $s\alpha$ for (s,α).

---

[1] To accommodate a fragment of English admitting tenses, one would add tense operators to the formalism of IL, as in Montague [1973]. The semantics of IL would then be extended to take account of the tense operators. See Cocchiarella [1966].

THE LOGIC IL 11

<u>Primitive Symbols</u>. For each $\alpha \in T$ we have a denumerable list of <u>variables</u>

$$x_\alpha^0, x_\alpha^1, x_\alpha^2, \ldots$$

and non-logical <u>constants</u>[2]

$$c_\alpha^0, c_\alpha^1, c_\alpha^2, \ldots$$

of type α, together with the improper symbols $\equiv$, λ, $\wedge$, $\vee$, [,]. We also denote the variables of type α, in their proper order, by

$$x_\alpha, y_\alpha, z_\alpha, u_\alpha, v_\alpha, w_\alpha, f_\alpha, g_\alpha, h_\alpha,$$
$$x'_\alpha, y'_\alpha, z'_\alpha, \ldots,$$

and the constants by

$$c_\alpha, d_\alpha, c'_\alpha, d'_\alpha, \ldots,$$

so that, e.g., h_α is x_α^8 and c''_α is c_α^4. We use the letters 'x', 'y', 'z', ... , 'h' (serif type), with or without superscripts or primes, as syntactical variables ranging over formal variables of IL, and similarly we use 'c', 'd', with or without superscripts or primes, to range over constants of IL.

<u>Terms</u>. We characterize recursively the set Tm_α of <u>terms of IL of type</u> α, as follows:

(i)   Every variable of type α belongs to Tm_α,

(ii)   Every constant of type α belongs to Tm_α,

(iii) $A \in Tm_{\alpha\beta}$, $B \in Tm_\alpha$ imply $[AB] \in Tm_\beta$,

(iv) $A \in Tm_\beta$, x a variable of type α imply $\lambda x\, A \in Tm_{\alpha\beta}$,

(v) $A, B \in Tm_\alpha$ imply $[A \equiv B] \in Tm_t$,

(vi) $A \in Tm_\alpha$ implies $\wedge A \in Tm_{s\alpha}$,

(vii) $A \in Tm_{s\alpha}$ implies $\vee A \in Tm_\alpha$.

---

[2] One could allow here an arbitrary set of constants, not necessarily denumerable. See comment at the end of §3.

We write A_α for A when $A \in Tm_\alpha$, and adopt the usual conventions regarding grouping in terms. In particular, we sometimes use parentheses (,) in place of brackets [,] , and outermost brackets may be omitted.

Semantics. Let D and I be non-empty sets. By the <u>standard frame based on</u> D <u>and</u> I we understand the indexed family $(M_\alpha)_{\alpha \in T}$ of sets, where

(i) $M_e = D$,

(ii) $M_t = 2 = \{0,1\}$,

(iii) $M_{\alpha\beta} = M_\beta^{M_\alpha} = \{F \mid F: M_\alpha \to M_\beta\}$

(iv) $M_{s\alpha} = M_\alpha^I = \{F \mid F: I \to M_\alpha\}$.

A <u>(standard) model of</u> IL <u>based on</u> D <u>and</u> I is a system $M = (M_\alpha, m)_{\alpha \in T}$, where

(i) $(M_\alpha)_{\alpha \in T}$ is the standard frame based on D and I ,

(ii) m (the <u>meaning</u> function) is a mapping which assigns to each constant c_α a function from I into M_α ; in symbols, $m(c_\alpha) \in M_\alpha^I$.

Intuitively, a constant c_e of type e represents a name, like 'Jones' or 'the morning star', and must therefore be assigned an individual concept, rather than an individual, as its meaning or intension. Similarly, a constant c_{et} of type et might represent a basic common noun, like 'thief', which must be assigned a property of individuals, rather than a set, as its meaning.

If M is a model based on D and I , the <u>domain</u> D of M is denoted by $Dom(M)$, and the <u>index set</u> I of M is denoted by $Ind(M)$. We denote by $As(M)$ the set of all <u>assignments over</u> M , i.e., all functions a on the set of variables of IL such that $a(x_\alpha) \in M_\alpha$ for every variable x_α of type α . If $a \in As(M)$, x_α is a variable of type α , and $X \in M_\alpha$ then $a(x/X)$ denotes the assignment a' whose value $a'(y)$ for a variable y is equal to X if y is x_α and $a(y)$ otherwise.

We define the <u>value</u> $V_{i,a}^M(A_\alpha)$ <u>in</u> M <u>of the term</u> A_α <u>with respect to the index</u> i <u>and the assignment</u> a , by the following recursion on the term A_α of IL (we suppress the superscript 'M'):

(1) $V_{i,a}(x_\alpha) = a(x_\alpha)$,

(2) $V_{i,a}(c_\alpha) = m(c_\alpha)(i)$,

(3) $V_{i,a}(A_{\alpha\beta} B_\alpha) = V_{i,a}(A_{\alpha\beta})[V_{i,a}(B_\alpha)]$,

(4) $V_{i,a}(\lambda x_\alpha A_\beta) =$ the function F on M_α whose value at $X \in M_\alpha$ is equal to $V_{i,a'}(A_\beta)$, where $a' = a(x/X)$,

(5) $V_{i,a}(A_\alpha \equiv B_\alpha) = 1$ if $V_{i,a}(A_\alpha) = V_{i,a}(B_\alpha)$, and 0 otherwise,

(6) $V_{i,a}(^\wedge A_\alpha) =$ the function F on I whose value at $j \in I$ is equal to $V_{j,a}(A_\alpha)$,

(7) $V_{i,a}(^\vee A_{s\alpha}) = V_{i,a}(A_{s\alpha})(i)$.

It is easily seen that we always have $V_{i,a}(A_\alpha) \in M_\alpha$. Here <u>value</u> plays the role of <u>extension</u> in our earlier discussion, although to allow for free variables both the extension and the intension of a term A_α will depend on the assignment a . Precisely, we define $\text{Ext}_a[A_\alpha, i]$ to be $V_{i,a}(A_\alpha)$, and $\text{Int}_a[A_\alpha]$ to be the function F on I whose value at $i \in I$ is $V_{i,a}(A_\alpha)$.[3]

Clauses (6) and (7) deserve special attention. The cap operator $^\wedge$ acts as a functional abstractor over indices, although s itself is not a type and no variables ranging over indices are present in IL.[4] Given a term A_α , an index i and an assignment a , we have $\text{Ext}_a[^\wedge A_\alpha, i] = \text{Int}_a[A_\alpha]$; i.e., for each term A_α of IL we can produce another term $^\wedge A_\alpha$ whose extension (with respect to a and any index i) is the intension of A_α (with respect to a). In particular, the extension $\text{Ext}_a[^\wedge A_\alpha, i]$ of $^\wedge A_\alpha$ is independent of $i \in I$, so that $\text{Int}_a[^\wedge A_\alpha]$ is always a constant function on I . The cup operator $^\vee$ is an inverse to the cap operator. Given a term A_{se} , for example, which denotes for a particular index and assign-

---

[3] We adopt this course for reasons of convenience. There would be no difficulty, however, in making the assignment part of the context of use, by replacing our present indices, i, by pairs (i,a).

[4] In Chapter 2 we consider a two-sorted theory of types which contains such variables.

ment a certain individual concept, the term $\check{}A$ is of type e and denotes the individual which is the value of the concept for the index in question. The terms A_α and $\check{}\hat{}A_\alpha$ will always have the same extensions, and hence also the same intensions. However, for an arbitrary term $A_{s\alpha}$, which may not be of the form $\hat{}B_\alpha$, the terms A and $\hat{}\check{}A$ need not have the same extensions.

An occurrence of a variable x_β in a term A_α is <u>bound</u> if it occurs within a part $\lambda x_\beta B_\gamma$, otherwise <u>free</u>. As usual, the value $V_{i,a}(A_\alpha)$ depends only on the values $a(x_\beta)$ for x_β free in A_α, so that, e.g., if A_α contains exactly the distinct variables x_β, y_γ free, and $X \in M_\beta$, $Y \in M_\gamma$, we can write

$$V_{i;X,Y}(A_\alpha)$$

to denote the value $V_{i,a}(A_\alpha)$ for any assignment a such that $a(x) = X$, $a(y) = Y$. In particular, if A_α is <u>closed</u>, i.e., contains no free variables, then $V_{i,a}(A_\alpha)$ is independent of the assignment a, and we write simply $V_i(A_\alpha)$.

The class MC of <u>modally closed</u> terms is the smallest class such that

(i) $x_\alpha \in$ MC for every variable x_α,
(ii) $\hat{}A_\alpha \in$ MC for every term A_α,
(iii) $[A_{\alpha\beta}B_\alpha] \in$ MC whenever $A_{\alpha\beta}$, $B_\alpha \in$ MC,
(iv) $[A_\alpha \equiv B_\alpha] \in$ MC whenever A_α, $B_\alpha \in$ MC,
(v) $\lambda x_\alpha A_\beta \in$ MC whenever $A_\beta \in$ MC.

It is easily checked that the value $V_{i,a}(A_\alpha)$ of a modally closed term A_α is independent of $i \in I$, and we can therefore write simply $V_a(A_\alpha)$. If A_α is both closed and modally closed we write $V(A_\alpha)$.

A <u>formula</u> of IL is a term A_t of type t. Given a model M, an index i and an assignment a, we say that the formula A is <u>satisfied in</u> M <u>by</u> i <u>and</u> a, and write

$$M, i, a \text{ sat } A,$$

if $V^M_{i,a}(A) = 1$. In the case that A is closed, modally closed, or both, we write respectively M, i sat A , M, a sat A , or M sat A . Also, we give the obvious meaning to an expression such as

  M; i; X,Y sat A ,

when A is a formula containing exactly the distinct free variables x_β , y_γ , and X , Y are elements of M_β , M_γ respectively. A formula A is <u>true in</u> M if M, i, a sat A for every index i and assignment a . A set Σ of formulas is <u>satisfied in</u> M <u>by</u> i <u>and</u> a , and we write M, i, a sat Σ , if M, i, a sat A for every A ∈ Σ . Σ is <u>satisfiable</u> in IL if M, i, a sat Σ for some model M , index i and assignment a . A formula A is a <u>semantical consequence</u>, in IL, of a set Γ of formulas, and we write

  Γ ⊨ A in IL,

if M, i, a sat A whenever M, i, a sat Γ . A formula A is <u>valid</u> in IL, and we write

  ⊨ A in IL,

if A is a semantical consequence in IL of the empty set of formulas, or equivalently, if A is true in every model of IL.

We introduce the sentential connectives, quantifiers and modal operators in IL by definition[5]:

$$T = [\lambda x_t\, x_t \equiv \lambda x_t\, x_t] ,$$
$$F = [\lambda x_t\, x_t \equiv \lambda x_t\, T] ,$$
$$\sim\, = \lambda x_t\, [F \equiv x_t] ,$$
$$\wedge\, = \lambda x_t\, \lambda y_t\, [\lambda f_{tt}\, [fx \equiv y] \equiv \lambda f_{tt}\, [f\, T]] ,$$
$$\rightarrow\, = \lambda x_t\, \lambda y_t\, [[x \wedge y] \equiv x] ,$$
$$\vee\, = \lambda x_t\, \lambda y_t\, [\sim x \rightarrow y] ,$$
$$\forall x_\alpha\, A = [\lambda x_\alpha\, A \equiv \lambda x_\alpha\, T] ,$$

---

[5] With the exception of the modal operators, these definitions follow Henkin [1963]. We write [A ∧ B] instead of [[∧ A] B] where A and B are formulas; similarly for the other binary connectives.

$$\exists x_\alpha A = \sim \forall x_\alpha \sim A ,$$

$$[A_\alpha \equiv B_\alpha] = [{}^\wedge A_\alpha \equiv {}^\wedge B_\alpha] ,$$

$$\Box A = [A \equiv T] ,$$

$$\Diamond A = \sim \Box \sim A .$$

It is readily verified that the connectives and quantifiers have their usual meanings in any model M under these definitions; hence, for example, M, i, a sat $[A \vee B]$ just in case either M, i, a sat A or M, i, a sat B, and M, i, a sat $\forall x_\alpha A$ just in case $M, i, a(x/X)$ sat A for every $X \in M_\alpha$. The necessity operator $\Box$ acts as a quantifier over indices: M, i, a sat $\Box A$ if and only if M, j, a sat A for every $j \in I$.[6]

Many of the usual principles of type theory -- tautologies, laws of rewrite for bound variables, etc. -- continue to hold in IL. However, there are some accustomed laws which turn out not to be valid in the intensional setting, in particular (as is typical of modal quantification theories) unrestricted laws of universal instantiation and substitutivity of equals. For example, the formulas

(i) $\quad \forall x_e \exists y_e [x_e \equiv y_e] \rightarrow \exists y_e [c_e \equiv y_e] ,$

(ii) $\quad c_e \equiv d_e \rightarrow [c_e \equiv c_e \rightarrow c_e \equiv d_e]$

are not valid in IL.[7] Various restricted formulations of these principles are valid in IL, however: Let $A(x_\alpha)$ be a term involving the variable x_α, and denote by $A(B_\alpha)$ the result of replacing all free occurrences of x_α in $A(x_\alpha)$ by the term B_α, rewriting bound variables in $A(x_\alpha)$ if necessary to avoid clashes. Then the following schemata are valid in IL:

(i) $\quad \forall x_\alpha A_t(x_\alpha) \rightarrow A_t(B_\alpha)$, provided no free occurrence of x_α lies in the scope of $^\wedge$ in $A_t(x_\alpha)$,

---

[6] Thus, necessity is an S5 operator; see Kripke [1963b].

[7] The failure of universal instantiation is due to the treatment of constants in IL. In the modal predicate logic of §9, which can be viewed as an alternative formulation of IL, an unrestricted principle of universal instantiation is valid. Cf. also the two-sorted type theory of §8.

(ii) $\forall x_\alpha A_t(x_\alpha) \to A_t(B_\alpha)$, if B_α is modally closed,

(iii) $B_\alpha \equiv C_\alpha \to A_\beta(B_\alpha) \equiv A_\beta(C_\alpha)$, provided no free occurrence of x_α lies in the scope of $\hat{}$ in $A_\beta(x_\alpha)$,

(iv) $B_\alpha \equiv C_\alpha \to A_\beta(B_\alpha) \equiv A_\beta(C_\alpha)$, if B_α and C_α are modally closed,

(v) $B_\alpha \equiv C_\alpha \to A_\beta(B_\alpha) \equiv A_\beta(C_\alpha)$.

§3. Generalized Completeness of IL

Since IL incorporates the theory of types, its valid formulas are not recursively enumerable, and therefore no complete axiomatization exists. In this section we prove a <u>generalized</u> completeness theorem for an axiomatic formulation of IL, based on the corresponding result in Henkin [1950].

<u>Generalized Semantics</u>.[1] Let D and I be non-empty sets. By a <u>frame based on</u> D <u>and</u> I we understand an indexed family $(M_\alpha)_{\alpha \in T}$ of sets, where

(i) $M_e = D$,

(ii) $M_t = 2 = \{0,1\}$,

(iii) $M_{\alpha\beta}$ is a non-empty subset of $M_\beta^{M_\alpha}$,

(iv) $M_{s\alpha}$ is a non-empty subset of M_α^I .

A <u>general model (g-model) of</u> IL <u>based on</u> D and I is a system $M = (M_\alpha, m)_{\alpha \in T}$, where

(i) $(M_\alpha)_{\alpha \in T}$ is a frame based on D and I ,

(ii) m (the <u>meaning</u> function) is a mapping which assigns to each constant c_α a function from I into M_α ,

---

[1] The terminology employed here is that of Henkin [1950].

(iii) There exists a function V^M (the <u>value</u> function) which assigns, to each $i \in I$, $a \in As(M)$, and $A_\alpha \in Tm_\alpha$, a value $V^M_{i,a}(A_\alpha) \in M_\alpha$, in such a way as to satisfy the recursive conditions (1) through (7) on page 13.

As remarked in Henkin [1950], this notion of general model is impredicative, as a result of clause (iii). The difficulty in attempting to <u>define</u> value in an arbitrary frame is caused by the recursive conditions (4) and (6) corresponding to λ and $\char94$, since the functions described there may simply fail to belong to the appropriate domain M_α. We must therefore add clause (iii), which stipulates the existence of V^M; the uniqueness of V^M follows immediately from the recursive conditions (1) through (7). In §6 we provide a more direct characterization of a large class of g-models.

The notion of <u>satisfaction</u> of a formula in a g-model is exactly as before, with <u>model</u> replaced by <u>g-model</u>. The formula A is a <u>g-semantical consequence</u>, in IL, of a set Γ of formulas, and we write

$\Gamma \models_g A$ in IL,

if M, i, a sat A whenever M is a g-model of IL, $i \in I$, $a \in As(M)$, and M, i, a sat Γ. If Γ is empty we say that A is <u>g-valid</u> in IL, and we write

$\models_g A$ in IL.

A set Σ of formulas is <u>g-satisfiable</u> in IL if M, i, a sat Σ for some g-model M, index i, and assignment a.

It should be remarked that the sentential connectives, quantifiers and modal operators continue to have their usual meaning even in general models. The valid schemata (i) through (v) given at the end of §2 continue to be g-valid, and since every standard model of IL is a g-model it is immediate that

$\Gamma \models_g A$ in IL implies $\Gamma \models A$ in IL,

$\models_g A$ in IL implies $\models A$ in IL,

Σ satisfiable in IL implies Σ g-satisfiable in IL.

The Theory IL. We give a deductive structure to the language of IL in the usual way, first specifying a recursive set of axioms and inference rules and then defining a theorem of IL to be any formula obtainable from the axioms by repeated application of the rules. We use the term IL to refer to both the language and this deductive theory within the language; no confusion should arise.

Axioms of IL.

A1. $g_{tt} T \wedge g_{tt} F \equiv \forall x_t [gx]$,

A2. $x_\alpha \equiv y_\alpha \rightarrow f_{\alpha t} x \equiv f_{\alpha t} y$,

A3. $\forall x_\alpha [f_{\alpha\beta} x \equiv g_{\alpha\beta} x] \equiv [f \equiv g]$,

AS4. $(\lambda x_\alpha A_\beta(x)) B_\alpha \equiv A_\beta(B_\alpha)$, where $A_\beta(B_\alpha)$ comes from $A_\beta(x_\alpha)$ by replacing all free occurrences of x by the term B , and (i) no free occurrence of x in $A(x)$ lies within a part $\lambda y\, C$ where y is free in B , and either (ii) no free occurrence of x in $A(x)$ lies within the scope of $\hat{}$, or else (ii') B is modally closed,

A5. $\Box [\check{}f_{s\alpha} \equiv \check{}g_{s\alpha}] \equiv [f \equiv g]$,

AS6. $\check{}\hat{}A_\alpha \equiv A_\alpha$.

Rule of Inference.

R. From $A_\alpha \equiv A'_\alpha$ and the formula B to infer the formula B' , where B' comes from B by replacing one occurrence of A (not immediately preceded by λ) by the term A' .

This axiomatization for IL corresponds very closely to the axiomatization for the theory of propositional types given in Henkin [1963], as simplified in Andrews [1963]. AS4 is just Henkin's Axiom Schema 7, suitably modified for IL. The new axioms A5 and AS6 are analogues of A3 and AS4, with the intensional abstractor $\hat{}$ playing the role of the functional abstractor λ . We remark in passing that AS4 can be replaced by the following schemata, corresponding to Henkin's schemata 7.1 through 7.5:

AS4.1 $(\lambda x_\alpha A_\beta) B_\alpha \equiv A_\beta$, if x_α is not free in A_β ,

AS4.2 $(\lambda x_\alpha x_\alpha) B_\alpha \equiv B_\alpha$,

AS4.3 $(\lambda x_\alpha [A_{\beta\gamma} C_\beta]) B_\alpha \equiv [(\lambda x A) B] [(\lambda x C) B]$,

AS4.4 $(\lambda x_\alpha [A_\beta \equiv C_\beta]) B_\alpha \equiv [(\lambda x A) B \equiv (\lambda x C) B]$,

AS4.5 $(\lambda x_\alpha \lambda y_\beta A_\gamma) B_\alpha \equiv \lambda y [(\lambda x A) B]$, if x and y are distinct and y is not free in B ,

AS4.6 $(\lambda x_\alpha {}^\vee A_{s\beta}) B_\alpha \equiv {}^\vee[(\lambda x A) B]$,

AS4.7 $(\lambda x_\alpha {}^\wedge A_\beta) B_\alpha \equiv {}^\wedge[(\lambda x A) B]$, if B is modally closed.

Similarly, Rule R can be replaced by the eight rules below:

R1. From $A_t \equiv A'_t$ and A to infer A' ,

R2. From $A_{\alpha\beta} \equiv A'_{\alpha\beta}$ to infer $A B_\alpha \equiv A' B_\alpha$,

R3. From $A_\alpha \equiv A'_\alpha$ to infer $B_{\alpha\beta} A \equiv B_{\alpha\beta} A'$,

R4. From $A_\beta \equiv A'_\beta$ to infer $\lambda x_\alpha A \equiv \lambda x_\alpha A'$,

R5. From $A_\alpha \equiv A'_\alpha$ to infer ${}^\wedge A \equiv {}^\wedge A'$,

R6. From $A_{s\alpha} \equiv A'_{s\alpha}$ to infer ${}^\vee A \equiv {}^\vee A'$,

R7. From $A_\alpha \equiv A'_\alpha$ to infer $[B_\alpha \equiv A] \equiv [B_\alpha \equiv A']$,

R8. From $A_\alpha \equiv A'_\alpha$ to infer $[A \equiv B_\alpha] \equiv [A' \equiv B_\alpha]$.

A *proof* in IL is a sequence of formulas each of which is either an axiom or else is obtainable from earlier formulas by Rule R. A formula A is *provable* in IL, or a *theorem* of IL, and we write

$\vdash A$ in IL,

if it is the last line of a proof in IL. If A is a formula and Γ is a set of formulas, we write

$\Gamma \vdash A$ in IL

and say that A is *derivable from* Γ in IL, if there exist formulas B^0, $B^1, \ldots, B^{n-1}$ in Γ such that

$B^0 \rightarrow . B^1 \rightarrow . \ldots \rightarrow . B^{n-1} \rightarrow A$

is a theorem of IL. A set Σ of formulas is *consistent* in IL if the formula F is not derivable from Σ in IL.

THEOREM 3.1 (Soundness Theorem for IL)

(i) $\vdash A$ in IL implies $\vDash_g A$ in IL,

(ii) $\Gamma \vdash A$ in IL implies $\Gamma \vDash_g A$ in IL,

(iii) Σ g-satisfiable in IL implies Σ consistent in IL.

Proof: The axioms of IL are easily seen to be g-valid (in the case of AS4 one can either show this directly or verify the g-validity of the equivalent AS4.1 through AS4.7), and Rule R clearly preserves g-validity. This proves (i), from which (ii) and (iii) follow immediately.

The Generalized Completeness Theorem for IL (Theorem 3.3) is the converse to Theorem 3.1. Before we can prove it we need some additional information about the theory IL, which is provided in the following list of metatheorems. The reader can easily reconstruct the proofs of these theorems, in the order listed, by consulting Henkin [1963] and Andrews [1963]; we remark that later theorems in the list frequently imply earlier ones (e.g., T35 implies T5.1), but in general these later theorems depend on earlier ones for their proof.

Some preliminary terminology: Given a term $A_\beta(x_\alpha)$ and a term B_α, we say that B is free for x in $A(x)$ if no free occurrence of x in $A(x)$ lies within a part $\lambda y\, C$, where y occurs free in B. A sentential formula is one built up from the formulas T, F and variables of type t by means of the connectives $\equiv$, $\sim$, $\wedge$, $\to$, $\vee$. A sentential formula is a tautology if it is valid (or equivalently, g-valid) in IL. An arbitrary formula of IL is tautologous if it comes from a tautology by uniform substitution of formulas of IL for free variables.

Metatheorems of IL.

T1. $\vdash A_\alpha \equiv A_\alpha$

T2. $\vdash T$

T3. $\vdash \forall x_\alpha T$

T4. $\vdash \Box T$

T5.1 $\vdash A_\alpha \equiv B_\alpha$ implies $\vdash B_\alpha \equiv A_\alpha$

T5.2 $\vdash A_\alpha \equiv B_\alpha$ and $\vdash B_\alpha \equiv C_\alpha$ imply $\vdash A_\alpha \equiv C_\alpha$

T6. $\vdash T \wedge T$

T7. $\vdash [A \equiv T] \equiv A$, where A is any formula.

T8. $\vdash A$ implies $\vdash \forall x_\alpha A$

T9. $\vdash \forall x_\alpha A(x)$ implies $\vdash A(B_\alpha)$, where $A(B)$ comes from the formula $A(x)$ by replacing all free occurrences of x by the term B, and (i) B is free for x in $A(x)$, and either (ii) no free occurrence of x in $A(x)$ lies within the scope of $\char94$, or else (ii') B is modally closed.

T10. $\vdash A(x_\alpha)$ implies $\vdash A(B_\alpha)$, where $A(x)$ and B satisfy the conditions of T9.

T11. $\vdash A$ and $\vdash B$ imply $\vdash A \wedge B$

T12. $\vdash \lambda x_\alpha A_\beta(x) \equiv \lambda y_\alpha A_\beta(y)$, where $A(x)$ and $A(y)$ are identical except that $A(x)$ has free occurrences of x where $A(y)$ has free occurrences of y, and vice-versa.

T13. $\vdash \forall x_\alpha A(x) \equiv \forall y_\alpha A(y)$, where $A(x)$ and $A(y)$ are formulas satisfying the conditions of T12.

T14. $\vdash \forall x_\alpha [A_{\alpha\beta} x \equiv B_{\alpha\beta} x] \equiv [A \equiv B]$, where x is any variable not occurring free in A or B.

T15. $\vdash A(T)$ and $\vdash A(F)$ imply $\vdash A(x_t)$, where $A(T)$ and $A(F)$ come from the formula $A(x)$ by replacing all free occurrences of x by T and F respectively.

T16. $\vdash T \wedge F \equiv F$

T17. $\vdash T \wedge x_t \equiv x_t$

T18. $\vdash [T \rightarrow x_t] \equiv x_t$

T19. $\vdash A \rightarrow B$ and $\vdash A$ imply $\vdash B$

T20. $\vdash x_\alpha \equiv y_\alpha \rightarrow f_{\alpha\beta} x \equiv f_{\alpha\beta} y$

T21. $\vdash f_{\alpha\beta} \equiv g_{\alpha\beta} \rightarrow fx_\alpha \equiv gx_\alpha$

T22. $\vdash F \rightarrow x_t$

T23. $\vdash A$ if A is a tautology

T24. $\vdash A$ if A is tautologous.

T25. $\vdash \forall x_\alpha A(x) \to A(B_\alpha)$, where $A(x)$ and B satisfy the conditions of T9.

T26. $\vdash A(B_\alpha) \to \exists x_\alpha A(x)$, under the same conditions.

T27. $\vdash \exists x_t\, x_t$

T28. $\vdash \exists x_t \sim x_t$

T29. $\vdash \exists x_\alpha [A_\alpha \equiv x]$, where x is any variable not free in A.

T30. $\vdash \forall x_\alpha [A \to B] \to [A \to \forall x B]$, where x is any variable not free in A.

T31. $\vdash B_\alpha \equiv C_\alpha \to A_\beta(B) \equiv A_\beta(C)$, where $A(B)$, $A(C)$ come from $A(x_\alpha)$ by replacing all free occurrences of x by the terms B, C respectively, and (i) B and C are free for x in $A(x)$, and either (ii) no free occurrence of x in $A(x)$ lies within the scope of $\hat{}$, or else (ii') B and C are modally closed.

T32. $\vdash B_\alpha \equiv B'_\alpha \to A_{\alpha\beta} B \equiv A_{\alpha\beta} B'$

T33. $\vdash A_{\alpha\beta} \equiv A'_{\alpha\beta} \to A B_\alpha \equiv A' B_\alpha$

T34. $\vdash A_{s\alpha} \equiv A'_{s\alpha} \to \check{\,}A \equiv \check{\,}A'$

T35. $\vdash A_\alpha \equiv B_\alpha \to B \equiv A$

T36. $\vdash A_\alpha \equiv B_\alpha \to [B_\alpha \equiv C_\alpha \to A \equiv C]$

T37. $\vdash B_\alpha \equiv C_\alpha \to A_\beta(B) \equiv A_\beta(C)$, where $A(B)$, $A(C)$ come from $A(x_\alpha)$ by replacing all free occurrences of x by the terms B, C respectively, and B and C are free for x in $A(x)$.

T38. $\vdash B_\alpha \equiv C_\alpha \to \Box [B \equiv C]$

T39. $\vdash A \to \Box A$, if A is a modally closed formula.

T40. $\vdash \Box A \to A$, where A is any formula.

T41. $\vdash A \to \Diamond A$

T42. $\vdash \sim \Box A \to \Box \sim \Box A$

T43. $\vdash \Box [A \to B] \to [\Box A \to \Box B]$

T44. $\vdash A$ implies $\vdash \Box A$

T45. $\vdash \sim \Box A \equiv \Diamond \sim A$

T46. $\vdash \sim \Diamond A \equiv \Box \sim A$

T47. $\vdash \Box \Diamond A \equiv \Diamond A$

T48. $\vdash \Box \Box A \equiv \Box A$

T49. $\vdash \Diamond \Box A \equiv \Box A$

T50. $\vdash \Diamond \Diamond A \equiv \Diamond A$

T51. $\vdash \Box [A \wedge B] \equiv \Box A \wedge \Box B$

T52. $\vdash \Diamond [A \vee B] \equiv \Diamond A \vee \Diamond B$

T53. $\vdash \Box A \vee \Box B \rightarrow \Box [A \vee B]$

T54. $\vdash \Diamond [A \wedge B] \rightarrow \Diamond A \wedge \Diamond B$

T55. $\vdash \Box [A \rightarrow B] \rightarrow [\Diamond A \rightarrow \Diamond B]$

T56. $\vdash \Box \forall x_\alpha A \equiv \forall x \Box A$

T57. $\vdash \Diamond \exists x_\alpha A \equiv \exists x \Diamond A$

T58. $\vdash \exists x_\alpha \Box A \rightarrow \Box \exists x A$

T59. $\vdash \Diamond \forall x_\alpha A \rightarrow \forall x \Diamond A$

T60. $\vdash \Box [\check{f}_{s\alpha} \equiv \check{g}_{s\alpha}] \equiv [f \equiv g]$

T61. $\vdash \Box [B_\alpha \equiv C_\alpha] \rightarrow B \equiv C$

T62. $\vdash \hat{A}_\alpha \equiv f_{s\alpha} \rightarrow \Box [A \equiv \check{f}]$

T63. $A \in \Gamma$ implies $\Gamma \vdash A$

T64. $\Gamma \vdash A$ and $\Gamma \subseteq \Delta$ imply $\Delta \vdash A$

T65. $\vdash A$ implies $\Gamma \vdash A$

T66. $\Gamma \vdash A$ and $\Gamma \vdash [A \rightarrow B]$ imply $\Gamma \vdash B$

T67. $\Gamma \vdash A$ implies $\Gamma \vdash \forall x_\alpha A$, where x is any variable not occurring free in Γ.

T68. $\Gamma \cup \{A\} \vdash B$ if and only if $\Gamma \vdash [A \rightarrow B]$

T69. $\Gamma \vdash A$ if and only if $\Gamma \cup \{\sim A\}$ is inconsistent.

We can now prove a key lemma needed for the Generalized Completeness Theorem for IL:

LEMMA 3.2. Let Σ be a set of formulas, consistent in IL, and suppose there are infinitely many variables of each type which do not occur in any formula in Σ. Then there exists a sequence $\overline{\Sigma} = (\overline{\Sigma}_i)_{i \in \omega}$ of sets of formulas such that:

(i) $\Sigma \subseteq \overline{\Sigma}_0$,

(ii) For each $i \in \omega$, $\overline{\Sigma}_i$ is a maximal consistent set of formulas in IL,

(iii) For each $i \in \omega$ and each formula $B(x_\alpha)$, we have $\exists x\, B(x) \in \overline{\Sigma}_i$ if and only if $B(y_\alpha) \in \overline{\Sigma}_i$ for some variable y which is free for x in $B(x)$,

(iv) For each $i \in \omega$ and each formula B, we have $\Diamond B \in \overline{\Sigma}_i$ if and only if $B \in \overline{\Sigma}_j$ for some $j \in \omega$.

Proof: If Δ is a finite set of formulas, let $\mathrm{Cnj}(\Delta)$ denote the conjunction, in some standard order, of the formulas of Δ, which we take to be the formula $\top$ if Δ is empty. Given an arbitrary sequence $\Gamma = (\Gamma_i)_{i \in \omega}$ of sets of formulas, we say that Γ is <u>relatively consistent</u> in IL if whenever $\Delta_i \subseteq \Gamma_i$ for all $i \in \omega$, each Δ_i finite, the set

$$\{ \Diamond \mathrm{Cnj}(\Delta_i) \mid i \in \omega \}$$

of formulas is consistent in IL. Given a formula A and $i \in \omega$, we say that A is <u>relatively i-consistent with</u> Γ in IL if the sequence obtained from Γ by adding A to Γ_i is relatively consistent in IL.

Let (i_0, A^0), (i_1, A^1), ... be an enumeration of all pairs (i, A), where $i \in \omega$ and A is a formula of IL. We define for each $k \in \omega$ a sequence $\Sigma^k = (\Sigma^k_i)_{i \in \omega}$ of sets of formulas, satisfying the conditions:

(1) For each $k \in \omega$, $\Sigma^k_i = \phi$ (the empty set) for all but finitely many values of $i \in \omega$,

(2) For each $k \in \omega$, there are infinitely many variables of each type which do not occur in any formula in any Σ^k_i for $i \in \omega$.

For $k = 0$ we put $\Sigma^0_0 = \Sigma$, $\Sigma^0_i = \phi$ for $i > 0$.

Given Σ^k satisfying (1) and (2), we define Σ^{k+1} as follows: Suppose (i_k, A^k) is the pair (i,A).

<u>Case 1</u>. A is not relatively i-consistent with Σ^k. Then take Σ^{k+1} to be Σ^k.

<u>Case 2</u>. A is relatively i-consistent with Σ^k.

<u>Case 2.1</u>. A does not have the form $\exists x_\alpha B$ or $\Diamond B$. Then let Σ^{k+1} be the same as Σ^k, except put $\Sigma_i^{k+1} = \Sigma_i^k \cup \{A\}$.

<u>Case 2.2</u>. A is $\exists x_\alpha B(x)$. Then let Σ^{k+1} be the same as Σ^k, except put $\Sigma_i^{k+1} = \Sigma_i^k \cup \{A, B(y_\alpha)\}$, where y is the first variable of type α which does not occur in Σ^k or $B(x)$, and $B(y)$ comes from $B(x)$ by replacing all free occurrences of x by y.

<u>Case 2.3</u>. A is $\Diamond B$. Then let Σ^{k+1} be the same as Σ^k, except put $\Sigma_i^{k+1} = \Sigma_i^k \cup \{A\}$, and put $\Sigma_j^{k+1} = \{B\}$, where j is the smallest number different from i for which $\Sigma_j^k = \phi$.

LEMMA 3.2.1. Each Σ^k is relatively consistent in IL.

<u>Proof</u>: By induction on k. If Σ^0 were relatively inconsistent then for some formulas $B^0, B^1, \ldots, B^{n-1}$ in Σ the formula

$$\Diamond [B^0 \wedge B^1 \wedge \ldots \wedge B^{n-1}]$$

would be inconsistent in IL, and hence Σ itself would be inconsistent in IL by T41, contradicting the hypothesis of Lemma 3.2. Assume Σ^k relatively consistent, and let (i_k, A^k) be the pair (i,A). We show that Σ^{k+1} is relatively consistent:

<u>Case 1</u>. A is not relatively i-consistent with Σ^k. Then $\Sigma^{k+1} = \Sigma^k$ and is therefore relatively consistent.

<u>Case 2</u>. A is relatively i-consistent with Σ^k.

<u>Case 2.1</u>. A does not have the form $\exists x_\alpha B$ or $\Diamond B$. Then Σ^{k+1} is obtained from Σ^k by adding A to Σ_i^k, and is therefore relatively consistent by the assumption under Case 2.

Case 2.2. A is $\exists x_\alpha\, B(x)$. Let y_α be the first variable of type α not occurring in Σ^k or $B(x)$, and suppose that Σ^{k+1} were relatively inconsistent. Then for some finite sets $\Delta_j \subseteq \Sigma_j^k$, $j \in \omega$, the set

$$\{\,\Diamond\, \mathrm{Cnj}(\Delta_j) \mid j \neq i\,\} \cup \{\,\Diamond\,[\,\mathrm{Cnj}(\Delta_i) \wedge A \wedge B(y)\,]\,\}$$

would be inconsistent in IL. But then, letting Γ be the set $\{\Diamond\,\mathrm{Cnj}(\Delta_j) \mid j \neq i\,\}$ and using T13, T56, T67, T69 and others, we would have:

$\Gamma \vdash \Box \sim [\,\mathrm{Cnj}(\Delta_i) \wedge A \wedge B(y)\,]$,

$\Gamma \vdash \forall y\, \Box \sim [\,\mathrm{Cnj}(\Delta_i) \wedge A \wedge B(y)\,]$,

$\Gamma \vdash \Box\, \forall y \sim [\,\mathrm{Cnj}(\Delta_i) \wedge A \wedge B(y)\,]$,

$\Gamma \vdash \Box \sim \exists y\, [\,\mathrm{Cnj}(\Delta_i) \wedge A \wedge B(y)\,]$,

$\Gamma \vdash \Box \sim [\,\mathrm{Cnj}(\Delta_i) \wedge A \wedge \exists y\, B(y)\,]$,

$\Gamma \vdash \Box \sim [\,\mathrm{Cnj}(\Delta_i) \wedge A \wedge \exists x\, B(x)\,]$,

$\Gamma \vdash \Box \sim [\,\mathrm{Cnj}(\Delta_i) \wedge A\,]$,

so that by T69 again the set

$$\{\,\Diamond\,\mathrm{Cnj}(\Delta_j) \mid j \neq i\,\} \cup \{\,\Diamond\,[\,\mathrm{Cnj}(\Delta_i) \wedge A\,]\,\}$$

would be inconsistent in IL, contradicting the assumption under Case 2 that A is relatively i-consistent with Σ^k.

Case 2.3. A is $\Diamond B$. Let j be the smallest number different from i for which $\Sigma_j^k = \phi$, and suppose that Σ^{k+1} were relatively inconsistent. Then for some finite sets $\Delta_\ell \subseteq \Sigma_\ell^k$, $\ell \in \omega$, the set

$$\{\,\Diamond\,\mathrm{Cnj}(\Delta_\ell) \mid \ell \neq i,j\,\} \cup \{\,\Diamond\,[\,\mathrm{Cnj}(\Delta_i) \wedge A\,]\,\} \cup \{\,\Diamond\, B\,\}$$

would be inconsistent in IL. But by T50, T54 we have

$\vdash \Diamond\,[\,\mathrm{Cnj}(\Delta_i) \wedge A\,] \rightarrow \Diamond B$,

so that the set

$$\{\,\Diamond\,\mathrm{Cnj}(\Delta_\ell) \mid \ell \neq i,j\,\} \cup \{\,\Diamond\,[\,\mathrm{Cnj}(\Delta_i) \wedge A\,]\,\}$$

would be inconsistent in IL, contradicting the assumption under Case 2.

LEMMA 3.2.2. The sequence $\bar{\Sigma} = (\bar{\Sigma}_i)_{i \in \omega}$ is relatively consistent, where

$$\bar{\Sigma}_i = \bigcup_{k \in \omega} \Sigma_i^k .$$

Proof: Clearly $\Sigma_i^k \subseteq \Sigma_i^{k+1}$ for $k, i \in \omega$. The assertion now follows in the usual way, using Lemma 3.2.1.

LEMMA 3.2.3. $\bar{\Sigma}_i$ is consistent in IL for each $i \in \omega$.

Proof: Follows from Lemma 3.2.2.

LEMMA 3.2.4. If A is a formula, $i \in \omega$, and A is relatively i-consistent with $\bar{\Sigma}$, then $A \in \bar{\Sigma}_i$.

Proof: Choose k so that the pair (i,A) is (i_k, A^k). Clearly A is relatively i-consistent with Σ^k, so by construction $A \in \Sigma_i^{k+1} \subseteq \bar{\Sigma}_i$.

LEMMA 3.2.5. If A is a formula and $\bar{\Sigma}_i \vdash A$, then $A \in \bar{\Sigma}_i$.

Proof: Straightforward, using Lemmas 3.2.2 and 3.2.4.

LEMMA 3.2.6. $\bar{\Sigma}_i$ is a maximal consistent set in IL; i.e., for every formula A either $A \in \bar{\Sigma}_i$ or else $\sim A \in \bar{\Sigma}_i$.

Proof: Suppose neither of the formulas A, $\sim$A belongs to $\bar{\Sigma}_i$. Then by Lemma 3.2.4 neither formula is relatively i-consistent with $\bar{\Sigma}$. It follows that there exist, for $j \in \omega$, finite sets $\Delta'_j, \Delta''_j \subseteq \bar{\Sigma}_j$ such that

$$\{ \Diamond \mathrm{Cnj}(\Delta'_j) \mid j \neq i \} \cup \{ \Diamond [\mathrm{Cnj}(\Delta'_i) \wedge A] \}$$

and

$$\{ \Diamond \mathrm{Cnj}(\Delta''_j) \mid j \neq i \} \cup \{ \Diamond [\mathrm{Cnj}(\Delta''_i) \wedge \sim A] \}$$

are both inconsistent in IL. But then, letting $\Delta_j = \Delta'_j \cup \Delta''_j$ and using T54, we see that

$$\Gamma \vdash \Box \sim [\mathrm{Cnj}(\Delta_i) \wedge A],$$
$$\Gamma \vdash \Box \sim [\mathrm{Cnj}(\Delta_i) \wedge \sim A],$$

where $\Gamma = \{ \Diamond \mathrm{Cnj}(\Delta_j) \mid j \neq i \}$, and from this it easily follows that

$\Gamma \vdash \Box \sim Cnj(\Delta_i)$, so that $\{\Diamond Cnj(\Delta_j) \mid j \in \omega\}$ is inconsistent in IL, contradicting Lemma 3.2.2.

LEMMA 3.2.7. For each $i \in \omega$ and each formula $B(x_\alpha)$, we have $\exists x B(x) \in \overline{\Sigma}_i$ if and only if $B(y_\alpha) \in \overline{\Sigma}_i$ for some variable y which is free for x in $B(x)$.

Proof: The implication from right to left follows from Lemma 3.2.5 and T26. For the implication from left to right, let A be the formula $\exists x B(x)$ and suppose $A \in \overline{\Sigma}_i$. Let (i,A) be the pair (i_k, A^k); then A is relatively i-consistent with Σ^k, since it is relatively i-consistent with $\overline{\Sigma}$ by Lemma 3.2.2. Therefore by the construction of Σ^{k+1}, we have $B(y_\alpha) \in \Sigma_i^{k+1} \subseteq \overline{\Sigma}_i$ for some variable y free for x in $B(x)$.

LEMMA 3.2.8. For each $i \in \omega$ and each formula B, we have $\Diamond B \in \overline{\Sigma}_i$ if and only if $B \in \overline{\Sigma}_j$ for some $j \in \omega$.

Proof: For the implication from right to left, suppose $B \in \overline{\Sigma}_j$ but $\Diamond B \notin \overline{\Sigma}_i$. Then $i \neq j$, in view of T41 and Lemma 3.2.5, and by Lemma 3.2.6 we have $\sim \Diamond B \in \overline{\Sigma}_i$ and therefore $\Box \sim B \in \overline{\Sigma}_i$ by T46. But then $\overline{\Sigma}$ is relatively inconsistent, since the set $\{\Diamond \Box \sim B, \Diamond B\}$ is inconsistent in IL by T49. This contradicts Lemma 3.2.2. The implication from left to right follows as in the proof of Lemma 3.2.7.

This completes the proof of Lemma 3.2.

REMARK 3.2.9. Suppose $\overline{\Sigma} = (\overline{\Sigma}_i)_{i \in \omega}$ satisfies (i) through (iv) of Lemma 3.2. Then it satisfies also:

(v) For each $i \in \omega$ and each formula $B(x_\alpha)$, we have $\forall x B(x) \in \overline{\Sigma}_i$ if and only if $B(y_\alpha) \in \overline{\Sigma}_i$ for every variable y which is free for x in $B(x)$.

(vi) For each $i \in \omega$ and each formula B, we have $\Box B \in \overline{\Sigma}_i$ if and only if $B \in \overline{\Sigma}_j$ for every $j \in \omega$.

We omit the proof, which is straightforward.

Lemma 3.2, it should be noted, depends for its proof only on the fact that the theorems of IL include the ordinary laws of sentential and predicate logic, together with the S5 modal laws T40, T42, T43 and T44. The lemma will therefore also hold for any theory in which these laws obtain, e.g. the first-order extensions of S5 described in Kripke [1959], Bayart [1958], or Hughes and Cresswell [1968]. Lemma 3.2 considerably simplifies the Henkin-type completeness proofs which have been given for these logics, and it is not difficult to modify the lemma and its proof to apply to quantified extensions of certain weaker modal logics than S5.[2]

We are now ready to prove:

THEOREM 3.3 (Generalized Completeness Theorem for IL)

(i) $\models_g$ A in IL implies $\vdash$ A in IL,

(ii) $\Gamma \models_g$ A in IL implies $\Gamma \vdash$ A in IL,

(iii) Σ consistent in IL implies Σ g-satisfiable in IL.

Proof: Parts (i) and (ii) follow easily from (iii) as usual, and in fact we show the somewhat stronger:

LEMMA 3.3.1. Suppose Σ is consistent in IL. Then Σ is g-satisfiable in a g-model $M = (M_\alpha, m)_{\alpha \in T}$ of IL based on sets D and I, where I is denumerable and D, as well as each domain M_α, is at most denumerable.

Proof: We may assume without loss of generality that there are infinitely many variables of each type not occurring in any formula of Σ, since, e.g., replacing each variable x_α^n by x_α^{2n} throughout Σ produces a set with this property, and clearly affects neither the consistency nor the g-satisfiability of Σ. Let $\bar{\Sigma} = (\bar{\Sigma}_i)_{i \in \omega}$ be a sequence of sets of formulas satisfying (i) through (iv) of Lemma 3.2, and therefore satisfying also (v) and (vi) of Remark 3.2.9.

Suppose $\alpha \in T$, $i \in \omega$. The relation

$$A_\alpha \simeq B_\alpha \pmod{i} \text{ if and only if } [A \equiv B] \in \bar{\Sigma}_i,$$

---

[2] In particular, to those extensions of the modal propositional logics K, M (also called T), B, and S4 (see Kaplan [1966]) in which the Barcan formula [∀x □A ↔ □∀x A] and the usual predicate laws are valid. Cf. the completeness proofs given in Cresswell [1967] and Thomason [1970].

between terms of type α, is clearly an equivalence relation on Tm_α, in view of T1, T35 and T36. Let us denote the set of variables of type α by Var_α. Then by T29 and property (iii) of $\overline{\Sigma}$, there exists for each $A \in Tm_\alpha$ and $i \in \omega$ some $x \in Var_\alpha$ for which $A \simeq x \pmod{i}$. In fact, property (iii) implies that there are infinitely many such variables x, since for any distinct $x^0, x^1, \ldots, x^{n-1} \in Var_\alpha$ the formula

$$\exists x \, [\, A \equiv x \wedge \forall x^0 \ldots \forall x^{n-1} \, [x \equiv x] \,]$$

is provable in IL, where x is not free in A and is distinct from x^0, $\ldots, x^{n-1}$. If $x, y \in Var_\alpha$ then by T39 the formula

$$x \equiv y \rightarrow \Box \, [x \equiv y]$$

is provable in IL, from which it follows that the relation $x \simeq y \pmod{i}$ is independent of $i \in \omega$ for variables x, y, and we can write simply $x \simeq y$. Let $x/\simeq$ denote the equivalence class of x_α in Var_α under this relation, and let D be the quotient set $Var_e/\simeq$ consisting of all classes $x/\simeq$ for $x \in Var_e$. Then D is at most denumerable. Letting $I = \omega$, we use the sequence $\overline{\Sigma}$ to construct, in the manner of Henkin [1950], a g-model $M = (M_\alpha, m)_{\alpha \in T}$ of IL based on D and I.

By recursion on $\alpha \in T$, we simultaneously define a set M_α and a mapping μ_α from Tm_α into M_α^I, satisfying the following three conditions:

(1)   For $i, j \in I$, $x \in Var_\alpha$: $\mu_\alpha(x)(i) = \mu_\alpha(x)(j)$,

(2)   For every $X \in M_\alpha$ there exists $x \in Var_\alpha$ such that $X = \mu_\alpha(x)(i)$ (which by (1) is independent of $i \in I$),

(3)   For $i \in I$, $A, B \in Tm_\alpha$: $\mu_\alpha(A)(i) = \mu_\alpha(B)(i)$ if and only if $A \simeq B \pmod{i}$.

$\underline{\alpha = t}$: Let $M_t = 2 = \{0,1\}$, and put $\mu_t(A_t)(i) = 1$ if $A_t \in \overline{\Sigma}_i$ and 0 otherwise. Then (1) follows from the fact that the formula $x_t \rightarrow \Box \, x_t$ is provable in IL, by T39. To verify (2), observe that by T27, T28 and property (iii) of $\overline{\Sigma}$, we must have $y \in \overline{\Sigma}_i$, $\sim z \in \overline{\Sigma}_i$ for some variables y, z of type t. Hence $z \notin \overline{\Sigma}_i$, and $\mu_t(y)(i) = 1$, $\mu_t(z)(i) = 0$. Condition (3) follows easily from the maximal consistency of $\overline{\Sigma}_i$.

32   INTENSIONAL LOGIC

$\underline{\alpha = e}$: Let $M_e = D = Var_e/\simeq$, and define $\mu_e(A_e)(i)$ to be $x/\simeq$, where x is any variable of type e for which $A \simeq x \pmod{i}$ (this is clearly well-defined). In particular, $\mu_e(x_e)(i) = x/\simeq$, so that clearly (1) and (2) hold. The verification of (3) is straightforward.

$\underline{\alpha = \beta\gamma}$: Suppose M_β , μ_β , M_γ , μ_γ have already been defined, and conditions (1), (2) and (3) hold for β and γ . We first define the mapping μ_α from Tm_α into $\left(M_\gamma^{M_\beta}\right)^I$, as follows: Given $A \in Tm_\alpha$, $i \in I$, and $X \in M_\beta$, let $x \in Var_\beta$ be chosen such that $X = \mu_\beta(x)(i)$, and put $\mu_\alpha(A)(i)(X) = \mu_\gamma(Ax)(i)$. Such a variable x exists by condition (2) for β ; to see that the value is well-defined, suppose that $y \in Var_\beta$ and $\mu_\beta(y)(i) = X = \mu_\beta(x)(i)$. Then by condition (3) for β , $x \simeq y \pmod{i}$, whence by T32 $Ax \simeq Ay \pmod{i}$ and therefore $\mu_\gamma(Ax)(i) = \mu_\gamma(Ay)(i)$ by condition (3) for γ .

Before defining M_α we verify condition (1) for α : Suppose $f \in Var_\alpha$ and $i, j \in I$; we show that $\mu_\alpha(f)(i) = \mu_\alpha(f)(j)$. Suppose $X \in M_\beta$. By conditions (1) and (2) for β we have $X = \mu_\beta(x)(i) = \mu_\beta(x)(j)$ for some $x \in Var_\beta$, and therefore $\mu_\alpha(f)(i)(X) = \mu_\gamma(fx)(i)$ and $\mu_\alpha(f)(j)(X) = \mu_\gamma(fx)(j)$, so it suffices to show that $\mu_\gamma(fx)(i) = \mu_\gamma(fx)(j)$. Let $y \in Var_\gamma$ be such that $fx \simeq y \pmod{i}$. By T39 the formula

$$fx \equiv y \to \Box[fx \equiv y]$$

is provable, so that $fx \simeq y \pmod{j}$ also, and therefore using conditions (1) and (3) for γ we have $\mu_\gamma(fx)(i) = \mu_\gamma(y)(i) = \mu_\gamma(y)(j) = \mu_\gamma(fx)(j)$.

We also observe that if $A \in Tm_\alpha$, $i \in I$ then $\mu_\alpha(A)(i) = \mu_\alpha(f)(i)$ for some $f \in Var_\alpha$. Indeed, suppose $A \simeq f \pmod{i}$; then if $X \in M_\beta$, say $X = \mu_\beta(x_\beta)(i)$, we have by T33 $Ax \simeq fx \pmod{i}$, so that by condition (3) for γ , $\mu_\alpha(A)(i)(X) = \mu_\gamma(Ax)(i) = \mu_\gamma(fx)(i) = \mu_\alpha(f)(i)(X)$. We can therefore set

$$M_\alpha = \{\mu_\alpha(f)(i) \mid f \in Var_\alpha\} \subseteq M_\gamma^{M_\beta} ,$$

which by condition (1) is independent of $i \in I$, and condition (2) will be satisfied for α . To verify condition (3), suppose $A, B \in Tm_\alpha$, $i \in I$. Then the following conditions are equivalent:

(a) $\mu_\alpha(A)(i) = \mu_\alpha(B)(i)$,

(b)   For every $Y \in M_\beta$, $\mu_\alpha(A)(i)(Y) = \mu_\alpha(B)(i)(Y)$,

(c)   For every $y \in Var_\beta$, $\mu_\alpha(A)(i)[\mu_\beta(y)(i)] = \mu_\alpha(B)(i)[\mu_\beta(y)(i)]$,

(d)   For every $y \in Var_\beta$, $\mu_\gamma(Ay)(i) = \mu_\gamma(By)(i)$,

(e)   For every $y \in Var_\beta$, $Ay \simeq By \pmod i$,

(f)   For every $y \in Var_\beta$, $[Ay \equiv By] \in \overline{\Sigma}_i$,

(g) $\forall x\ [Ax \equiv Bx] \in \overline{\Sigma}_i$, where x is the first variable of type β not occurring free in A or B ,

(h) $[A \equiv B] \in \overline{\Sigma}_i$,

(i) $A \simeq B \pmod i$.

These equivalences employ condition (2) for β , condition (3) for γ , property (v) of $\overline{\Sigma}$, and T14.

We remark that $A \in Tm_\alpha$, $B \in Tm_\beta$, $i \in I$ imply that

$$\mu_\gamma(AB)(i) = \mu_\alpha(A)(i)[\mu_\beta(B)(i)] .$$

For, suppose $B \simeq x_\beta \pmod i$. Then by T32 we have $AB \simeq Ax \pmod i$, from which $\mu_\gamma(AB)(i) = \mu_\gamma(Ax)(i) = \mu_\alpha(A)(i)[\mu_\beta(x)(i)] = \mu_\alpha(A)(i)[\mu_\beta(B)(i)]$.

$\underline{\alpha = s\beta}$: Suppose M_β , μ_β have already been defined so that conditions (1), (2) and (3) hold for β . We first define the mapping μ_α from Tm_α into $[M_\beta^I]^I$, as follows: Given $A \in Tm_\alpha$, $i \in I$, let $f \in Var_\alpha$ be chosen so that $A \simeq f \pmod i$, and put $\mu_\alpha(A)(i) = \mu_\beta(\check{f}) \in M_\beta^I$. This is well-defined, for if we also have $g \in Var_\alpha$ and $A \simeq g \pmod i$, then $f \simeq g$, i.e., $f \simeq g \pmod j$ for all $j \in I$, and therefore $\check{f} \simeq \check{g} \pmod j$ for all $j \in I$ by T34. Using condition (3) for β , this implies that $\mu_\beta(\check{f})(j) = \mu_\beta(\check{g})(j)$ for all $j \in I$, so that $\mu_\beta(\check{f}) = \mu_\beta(\check{g})$ in M_β^I .

We observe that for $f \in Var_\alpha$, $i \in I$, we have $\mu_\alpha(f)(i) = \mu_\beta(\check{f})$, which is independent of $i \in I$; thus condition (1) holds for α . Also, as earlier, if $A \in Tm_\alpha$, $i \in I$, then $\mu_\alpha(A)(i) = \mu_\alpha(f)(i)$ for some $f \in Var_\alpha$, since $A \simeq f \pmod i$ implies $\mu_\alpha(A)(i) = \mu_\beta(\check{f}) = \mu_\alpha(f)(i)$.

We can therefore set

$$M_\alpha = \{ \mu_\alpha(f)(i) \mid f \in \text{Var}_\alpha \} \subseteq M_\beta^I ,$$

which by condition (1) is independent of $i \in I$, and condition (2) will be satisfied for α. Finally, we show that condition (3) holds for α: Suppose $A, B \in \text{Tm}_\alpha$, $i \in I$. Choose $f, g \in \text{Var}_\alpha$ with $A \simeq f \pmod{i}$, $B \simeq g \pmod{i}$. Then the following conditions are easily seen to be equivalent, using condition (3) for β, property (vi) of $\overline{\Sigma}$, and T60:

(a) $\mu_\alpha(A)(i) = \mu_\alpha(B)(i)$,

(b) $\mu_\beta(\check{}f) = \mu_\beta(\check{}g)$,

(c) For all $j \in I$, $\mu_\beta(\check{}f)(j) = \mu_\beta(\check{}g)(j)$,

(d) For all $j \in I$, $\check{}f \simeq \check{}g \pmod{j}$,

(e) For all $j \in I$, $[\check{}f \equiv \check{}g] \in \overline{\Sigma}_j$,

(f) $\square [\check{}f \equiv \check{}g] \in \overline{\Sigma}_i$,

(g) $[f \equiv g] \in \overline{\Sigma}_i$,

(h) $f \simeq g \pmod{i}$,

(i) $A \simeq B \pmod{i}$.

This completes the definition of the frame $(M_\alpha)_{\alpha \in T}$ for IL based on D and I. Obviously, by conditions (1) and (2) each domain M_α is at most denumerable. To complete the definition of the g-model $M = (M_\alpha, m)_{\alpha \in T}$ we define the meaning function m by putting

$$m(c) = \mu_\alpha(c) \in M_\alpha^I$$

for each constant c_α.

It remains to prove that there exists a value function V^M in M. Suppose $A \in \text{Tm}_\alpha$, $a \in \text{As}(M)$. Suppose the free variables of A are among the distinct variables $x^0, x^1, \ldots, x^{n-1}$, where x^k is of type α_k, and write $A(x^0, \ldots, x^{n-1})$ for A. Choose a sequence $y^0, y^1, \ldots, y^{n-1}$ of distinct variables, y^k of type α_k, satisfying the conditions

(a) $\mu_{\alpha_k}(y^k)(i) = a(x^k)$ (independent of $i \in I$),

(b) y^k is free for x^k in A .

Such a sequence exists; for by conditions (1) and (2) for a_k , $a(x^k) = \mu_{\alpha_k}(y)(i)$, independent of $i \in I$, for some variable y of type α_k , and as remarked earlier we have $y \simeq y'$ (and therefore $\mu(y')(i) = \mu(y)(i) = a(x^k)$) for infinitely many variables y' . Hence for each $k < n$ there exist infinitely many variables y^k satisfying (a), and it follows that there exists a sequence $y^0, y^1, \ldots, y^{n-1}$ of distinct variables satisfying both (a) and (b). We call such a sequence a <u>representing sequence</u> for the term A and assignment a , where the original sequence $x^0, x^1, \ldots, x^{n-1}$ is fixed in advance to contain all free variables of A . Let $\bar{A}$ be the term $A(y^0, \ldots, y^{n-1})$ which results from A by simultaneously replacing all free occurrences of x^k by y^k for $k < n$. Given $i \in I$ we define

$$V^M_{i,a}(A) = \mu_\alpha(\bar{A})(i) \in M_\alpha .$$

We show that this is well-defined, i.e., the value $\mu_\alpha(\bar{A})(i)$ does not depend on the sequence $y^0, \ldots, y^{n-1}$. Indeed, suppose $z^0, \ldots, z^{n-1}$ is another representing sequence for A and a ; then $\mu_{\alpha_k}(y^k)(i) = a(x^k) = \mu_{\alpha_k}(z^k)(i)$, so we have $y^k \simeq z^k$ (mod i) , from which it follows by n applications of T31 that $A(y^0, \ldots, y^{n-1}) \simeq A(z^0, \ldots, z^{n-1})$ (mod i) , and therefore $\mu_\alpha[A(y^0, \ldots, y^{n-1})](i) = \mu_\alpha[A(z^0, \ldots, z^{n-1})](i)$, as required.

<u>Claim</u>: V^M is a value function in M .

We must verify the recursive clauses (1) through (7) on page 13. The verification of (1) and (2) is immediate. To verify

(3) $V_{i,a}(A_{\alpha\beta} B_\alpha) = V_{i,a}(A_{\alpha\beta})[V_{i,a}(B_\alpha)]$,

we let $x^0, \ldots, x^{n-1}$ be the distinct free variables of $[AB]$, and we choose a representing sequence $y^0, \ldots, y^{n-1}$ for $[AB]$ and a . We then have $V_{i,a}(AB) = \mu_\beta(\overline{AB})(i) = \mu_\beta(\bar{A}\bar{B})(i) = \mu_{\alpha\beta}(\bar{A})(i)[\mu_\alpha(\bar{B})(i)] = V_{i,a}(A)[V_{i,a}(B)]$.

(4) $V_{i,a}(\lambda x_\alpha\, A_\beta) =$ the function F on M_α whose value at $Y \in M_\alpha$ is equal to $V_{i,a'}(A_\beta)$, where $a' = a(x/Y)$.

For, let $Y \in M_\alpha$, $a' = a(x/Y)$; we show that $V_{i,a}(\lambda x\, A)(Y) = V_{i,a'}(A)$. Let $x, x^0, x^1, \ldots, x^{n-1}$ be a list of distinct variables which includes all free variables of A, and write $A(x, x^0, \ldots, x^{n-1})$ for A. Choose a representing sequence $y^0, \ldots, y^{n-1}$ for $\lambda x\, A$ and a, where of course y^k corresponds to x^k. Let y_α be a variable distinct from each y^k, free for x in A, and satisfying $\mu_\alpha(y)(i) = Y$. Then the sequence y, $y^0, \ldots, y^{n-1}$ forms a representing sequence for A and a', and we therefore have $V_{i,a'}(A) = \mu_\beta(\overline{A})(i)$, where $\overline{A}$ is $A(y, y^0, \ldots, y^{n-1})$, and also $V_{i,a}(\lambda x\, A)(Y) = \mu_{\alpha\beta}(\overline{\lambda x\, A})(i)(Y) = \mu_{\alpha\beta}(\overline{\lambda x\, A})(i)[\mu_\alpha(y)(i)] = \mu_\beta[(\overline{\lambda x\, A})y](i)$, where $\overline{\lambda x\, A}$ is the term $\lambda x\, A(x, y^0, \ldots, y^{n-1})$. It therefore suffices to show that $\mu_\beta[(\overline{\lambda x\, A})y](i) = \mu_\beta(\overline{A})(i)$, or equivalently that $(\lambda x\, A(x, y^0, \ldots, y^{n-1}))y \simeq A(y, y^0, \ldots, y^{n-1})$ (mod i). But this follows immediately from axiom AS4 of IL.

The verification of clause (5) is straightforward, similar to that for clause (3). To verify

(6) $V_{i,a}({}^\wedge A_\alpha) = $ the function F on I whose value at $j \in I$ is equal to $V_{j,a}(A_\alpha)$,

suppose that $j \in I$ and let $x^0, \ldots, x^{n-1}$ be the distinct free variables of A. Choose a representing sequence $y^0, \ldots, y^{n-1}$ for ${}^\wedge A$ and a. Then we have $V_{j,a}(A) = \mu_\alpha(\overline{A})(j)$, and in addition $V_{i,a}({}^\wedge A)(j) = \mu_{s\alpha}(\overline{{}^\wedge A})(i)(j) = \mu_{s\alpha}(\overline{{}^\wedge A})(i)(j) = \mu_\alpha({}^\vee f)(j)$, where $f \in \text{Var}_{s\alpha}$ is chosen so that ${}^\wedge\overline{A} \simeq f$ (mod i). It therefore suffices to show that $\mu_\alpha({}^\vee f)(j) = \mu_\alpha(\overline{A})(j)$, i.e., that $\overline{A} \simeq {}^\vee f$ (mod j). But this follows from T62 and property (vi) of $\overline{\Sigma}$.

(7) $V_{i,a}({}^\vee A_{s\alpha}) = V_{i,a}(A_{s\alpha})(i)$.

For, let $x^0, \ldots, x^{n-1}$ be the distinct free variables of A, and choose a representing sequence $y^0, \ldots, y^{n-1}$ for ${}^\vee A$ and a. Let $f \in \text{Var}_{s\alpha}$ be such that $\overline{A} \simeq f$ (mod i). Then $V_{i,a}(A)(i) = \mu_{s\alpha}(\overline{A})(i)(i) = \mu_\alpha({}^\vee f)(i)$ and $V_{i,a}({}^\vee A) = \mu_\alpha(\overline{{}^\vee A})(i) = \mu_\alpha({}^\vee \overline{A})(i)$, so it suffices to show that ${}^\vee \overline{A} \simeq {}^\vee f$ (mod i), which follows from T34.

This proves the claim, and therefore completes the proof that M is a g-model of IL. Now let $a \in \text{As}(M)$ be defined by:

$a(x_\alpha) = \mu_\alpha(x)(i)$ (independent of $i \in I$).

If A_α is any term, we clearly have $V_{i,a}(A) = \mu_\alpha(A)(i)$ for any $i \in I$, and this implies in particular that for any formula A of IL, M, i, a sat A if and only if $A \in \overline{\Sigma}_i$. Since $\Sigma \subseteq \overline{\Sigma}_0$, it follows that M, i, a sat Σ when $i = 0$. This completes the proof of Lemma 3.3.1 and Theorem 3.3.

Combining Theorems 3.1 and 3.3 we immediately obtain the following generalized compactness theorem for IL, a result which can also be proved directly using ultraproducts:

COROLLARY 3.4. Let Σ be a set of formulas of IL. Then Σ is g-satisfiable in IL if and only if every finite subset Σ' of Σ is g-satisfiable in IL.

We conclude this section by remarking that the entire development to this point admits a natural generalization to the case of a non-denumerable language; i.e., a formulation of IL which permits, for some cardinal κ, constants c_α^ξ for $\alpha \in T$ and $\xi < \kappa$.

§4. Persistence in IL

The Generalized Completeness Theorem proved in the last section relates the axiomatic theory IL to the generalized semantics for IL. However, we can also obtain from it a useful relationship between the theory IL and the standard semantics for IL described in §2. This relationship is expressed in Theorem 4.2, which shows that with respect to a suitably restricted class of formulas of IL, the theory IL is complete for the standard semantics.[1]

Suppose $M = (M_\alpha, m)_{\alpha \in T}$ is a g-model of IL based on D and I. Let $(M'_\alpha)_{\alpha \in T}$ be the standard frame based on D and I. For each $\alpha \in T$ we define an element $A'_\alpha \in M'_\alpha$:

---

[1] A somewhat weaker form of this result is anticipated in Montague [1970a], pp. 88-89.

(i) $\Lambda'_e \in D$, chosen arbitrarily,

(ii) $\Lambda'_t = 0$,

(iii) $\Lambda'_{\alpha\beta}(X') = \Lambda'_\beta$ for every $X' \in M'_\alpha$,

(iv) $\Lambda'_{s\alpha}(i) = \Lambda'_\alpha$ for every $i \in I$.

By recursion on $\alpha \in T$ we define a one-to-one mapping $\Phi_\alpha : M_\alpha \to M'_\alpha$:

(i) Φ_e is the identity mapping on $M_e = D = M'_e$,

(ii) Φ_t is the identity mapping on $M_t = 2 = M'_t$,

(iii) $\Phi_{\alpha\beta}(F)$, for $F \in M_{\alpha\beta}$, is the function $F' \in M'_{\alpha\beta}$ such that for all $X' \in M'_\alpha$, $F'(X') = \Phi_\beta[F(\Phi_\alpha^{-1}(X'))]$ if X' belongs to the range of Φ_α, and $F'(X') = \Lambda'_\beta$ otherwise,

(iv) $\Phi_{s\alpha}(F)$, for $F \in M_{s\alpha}$, is the function $F' \in M'_{s\alpha}$ such that for all $i \in I$, $F'(i) = \Phi_\alpha[F(i)]$.

Define a standard model $M' = (M'_\alpha, m')_{\alpha \in T}$ of IL based on D and I by letting $m'(c_\alpha)(i) = \Phi_\alpha[m(c_\alpha)(i)]$ for each constant c_α. A term A_α of IL is called M-<u>persistent</u> if, for every $i \in I$ and $a \in As(M)$, and for every choice of $\Lambda'_e \in D$, we have

$$\Phi_\alpha[V^M_{i,a}(A)] = V^{M'}_{i,a'}(A),$$

where $a' \in As(M')$ is defined by $a'(x_\alpha) = \Phi_\alpha[a(x_\alpha)]$ for every variable x_α. In particular, then, a formula A will be M-persistent if, for every $i \in I$ and $a \in As(M)$, and for every choice of $\Lambda'_e \in D$, the equivalence

M, i, a sat A if and only if M', i, a' sat A

holds. A term A_α is <u>persistent</u> if it is M-persistent for every g-model M of IL.

It is immediate from Theorem 3.1 that any term provably equal, in IL, to a persistent term is itself persistent. Also, a term A_α is persistent if and only if the formula $[A \equiv x]$ is persistent, where x is a variable of type α not occurring free in A.

The next theorem shows a large recursive class of terms to be persistent, and therefore provides a partial characterization of the class of all persistent terms.[2] A <u>truth-functional</u> type $\tau \in T$ is one which is built up from the type t alone by pairing; e.g., the type $(tt)(t(tt))$. If B, C^0, C^1, ..., C^{n-1} are terms of IL of types α, α_0, α_1, ..., α_{n-1} respectively, where

$$\alpha = \alpha_0(\alpha_1(\ldots (\alpha_{n-1} \beta)\ldots)) ,$$

then $BC^0C^1 \ldots C^{n-1}$ stands for the term $[\ldots[[BC^0]C^1]\ldots]C^{n-1}$ of type β, or B alone in the case $n = 0$.

THEOREM 4.1. Let Per denote the class of all persistent terms of IL. Then:

(i) All variables and constants belong to Per,

(ii) $A_{\alpha\beta}$, $B_\alpha \in$ Per imply $[AB] \in$ Per,

(iii) A_α, $B_\alpha \in$ Per imply $[A \equiv B] \in$ Per,

(iv) $A_\alpha \in$ Per implies $\hat{\ }A \in$ Per,

(v) $A_{s\alpha} \in$ Per implies $\check{\ }A \in$ Per,

(vi) A_t, $B_t \in$ Per imply $\sim A$, $[A \wedge B]$, $\square A \in$ Per,

(vii) $A_\alpha \in$ Per implies $\lambda x_e A \in$ Per,
      $A_t \in$ Per implies $\forall x_e A$, $\exists x_e A \in$ Per,

(viii) $A_\alpha \in$ Per implies $\lambda x_\tau A \in$ Per,
       $A_t \in$ Per implies $\forall x_\tau A$, $\exists x_\tau A \in$ Per (τ a truth-functional type)

(ix) Suppose $A_t \in$ Per and $F_t(x_\beta)$ is the formula $BC^0C^1 \ldots x \ldots C^n$, where C^m has type α_m for $m \leq n$, C^k is the term x_β, and B has type $\alpha_0(\alpha_1(\ldots (\alpha_n t)\ldots))$. Suppose also that x_β is not free in B, and the terms B, C^0, ..., C^n belong to Per. Then the formulas $\forall x [F(x) \rightarrow A]$ and $\exists x [F(x) \wedge A]$ belong to Per.

---

[2] This notion of persistence is a generalization to IL of a related notion for ordinary (non-modal) higher-order predicate logic. The question of a complete syntactic characterization for that case is discussed in Orey [1959]. See also Mostowski [1947].

(x) If A_t, $B_\alpha \in$ Per and x_α does not occur free in B_α, then the formula $\exists x\, [\, B \equiv x \wedge A\,]$ belongs to Per.

Proof: (i) through (vi) are straightforward. For (vii) and (viii) it suffices to observe that for a given g-model M of IL, if M' and the mappings Φ_α are obtained from M as earlier, then Φ_α maps M_α onto M'_α when α is e or a truth-functional type τ. In fact, for these types α it holds that $M_\alpha = M'_\alpha$ and Φ_α is just the identity mapping. For $\alpha = e$ this is obvious; for truth-functional types τ it follows easily from a result of Henkin.[3] Conditions (ix) and (x) are straightforward.

From the Generalized Completeness Theorem we now obtain:

THEOREM 4.2. Let Γ and Σ be sets of persistent formulas, A a persistent formula. Then:

(i) $\models$ A in IL if and only if $\vdash$ A in IL,

(ii) $\Gamma \models$ A in IL if and only if $\Gamma \vdash$ A in IL,

(iii) Σ is consistent in IL if and only if Σ is satisfiable in IL.

Proof: (iii) follows immediately from Theorem 3.3 (iii) and the definition of persistence. (i) and (ii) follow easily from (iii), as does the following

COROLLARY 4.3. Let Σ be a set of persistent formulas. Then Σ is satisfiable in IL if and only if every finite subset Σ' of Σ is satisfiable in IL.

Theorem 4.2 has application to various fragments of English, as described in §1. In particular, it is possible to show that Extensional English can be translated into IL in such a way that the translate of every sentence is a persistent closed formula of IL. This implies by Theorem 4.2 that a sentence of Extensional English will be valid if and only if its translate is provable in IL, and since the translation is effective it follows that the valid sentences of Extensional English are recursively enumerable. Moreover, Corollary 4.3 yields a compactness theorem for this fragment of English.

---

[3] Henkin [1963], §4.

CHAPTER 2. ALTERNATIVE FORMULATIONS OF IL

§5. Modal T-Logic

We take up now the first of several alternative formulations of the logic IL of Chapter 1. The most natural first step is to try to eliminate the functional and intensional abstractors λ and $\hat{}$ in favor of the more familiar quantifier $\forall$ and modal operator $\Box$, particularly in view of the fact that the abstractors λ, $\hat{}$ were responsible for the impredicativity of the notion of a general model of IL. For the moment, however, we choose to retain the full "functional" type structure, i.e., to allow variables and constants of all types $\alpha \in T$. The resulting logic we refer to as Modal T-Logic, and denote by ML_T. Since $\forall$ and $\Box$ are already defined in IL, the language of ML_T can be described as a sublanguage of the language of IL. We shall adopt this course, since it will facilitate a later comparison of the two logics.

Grammar. The atomic terms of ML_T comprise the smallest set ATm of terms of IL such that:

(i)   All variables and constants belong to ATm,

(ii) $A_{\alpha\beta}$, $B_\alpha \in$ ATm imply $[AB] \in$ ATm,

(iii) $A_{s\alpha} \in$ ATm implies $\check{}A \in$ ATm.

An example is the term $[\check{}f_{s(et)}[c_{ee}\check{}g_{se}]]$ of type t. A formula A of IL is an atomic formula of ML_T if one of the following holds:

(i)   A is an atomic term of ML_T of type t,

(ii)   A is $[B \equiv C]$, where B_α and C_α are atomic terms.

The formulas of ML_T are generated from the atomic formulas by the connectives $\sim$, $\wedge$, $\rightarrow$, $\vee$, the quantifier $\forall x_\alpha$ where $x_\alpha \in Var_\alpha$, and the necessity operator $\Box$. Thus, every formula of ML_T is also a formula of IL, but not conversely. Since $\exists$ and $\diamondsuit$ are defined in IL in terms of $\forall$, $\sim$ and $\Box$, these operators are also defined in ML_T.[1]

Generalized Semantics. Let D and I be non-empty sets. By a general model (g-model) of ML_T based on D and I we understand a system $M = (M_\alpha, m)_{\alpha \in T}$ such that:

(i) $(M_\alpha)_{\alpha \in T}$ is a frame based on D and I (see page 17),

(ii) m is a mapping which assigns to each constant c_α a function from I into M_α.

Thus, a g-model of ML_T differs from a g-model of IL only in that no value function V^M is assumed to exist in the former, and indeed none will exist in general. However, for formulas of ML_T the assumption that V^M exists is unnecessary. Specifically, let $M = (M_\alpha, m)_{\alpha \in T}$ be a g-model of ML_T based on D and I, and let $i \in I$, $a \in As(M)$. Then $V^M_{i,a}(A_\alpha) \in M_\alpha$ can be defined by recursion for every atomic term A_α in the usual way, and the notion

$$M, i, a \text{ sat } A,$$

where A is a formula of ML_T, can then be defined as follows:

(i) M, i, a sat A if and only if $V^M_{i,a}(A) = 1$, when A is an atomic term of type t,

(ii) M, i, a sat $[B \equiv C]$ if and only if $V^M_{i,a}(B) = V^M_{i,a}(C)$, when B and C are atomic terms,

(iii) Usual satisfaction clauses for $\sim$, $\wedge$, $\rightarrow$, $\vee$,

(iv) M, i, a sat $\forall x_\alpha A$ if and only if M, i, a(x/X) sat A for every $X \in M_\alpha$,

(v) M, i, a sat $\Box A$ if and only if M, j, a sat A for all $j \in I$.

---

[1] We could take $\sim$ and $\rightarrow$ as the only connectives in ML_T, defining the other connectives in terms of these two as usual. However, the formulas $[A \wedge B]$ and $[A \vee B]$ would then have two readings, one in IL and another in ML_T. This would unnecessarily complicate the later exposition.

This definition of satisfaction coincides with that given in §3 in the case when M is a g-model of IL and A is a formula of ML_T, so the same notation can be used without confusion.

Let A be either an atomic term or a formula of ML_T. We say A is <u>modally closed</u> if it is modally closed as a term of IL. For atomic terms or atomic formulas A, this holds just in case A contains no constants and no occurrence of $\vee$. The formula $\Box A$ is modally closed for any formula A, and the set of modally closed formulas of ML_T is closed under the connectives $\sim$, $\wedge$, $\rightarrow$, $\vee$ and the quantifier $\forall x_\alpha$. If A is a modally closed formula we can write M, a sat A instead of M, i, a sat A, as earlier. Corresponding to our generalized semantical notions for IL, we have for formulas of ML_T the notions $\Gamma \models_g A$ in ML_T, $\models_g A$ in ML_T, and Σ is g-satisfiable in ML_T, obtained from the corresponding notions in §3 (page 18) by replacing <u>g-model of</u> IL by <u>g-model of</u> ML_T throughout. It should be noted, however, that $\Gamma \models_g A$ in IL does not imply $\Gamma \models_g A$ in ML_T, even for formulas of ML_T.

<u>The Theory</u> ML_T. We can give an intrinsic axiomatization for ML_T, as follows:

<u>Axioms of</u> ML_T.

AS1.   A, where A is tautologous in $\sim$, $\wedge$, $\rightarrow$, $\vee$,

A2. $x_t \rightarrow [y_t \rightarrow x \equiv y]$,

A3. $\sim x_t \rightarrow [\sim y_t \rightarrow x \equiv y]$,

AS4. $\forall x_\alpha [A \rightarrow B] \rightarrow [A \rightarrow \forall x B]$, where x is any variable not occurring free in the formula A,

AS5. $\forall x_\alpha A(x) \rightarrow A(B_\alpha)$, where $A(B)$ comes from the formula $A(x)$ by replacing all free occurrences of x by the atomic term B, and (i) B is free for x in $A(x)$, and either (ii) no free occurrence of x in $A(x)$ lies within the scope of $\Box$, or else (ii') B is modally closed,

AS6. $\forall x_\alpha [A_{\alpha\beta} x \equiv B_{\alpha\beta} x] \rightarrow A \equiv B$, where x is any variable not occurring free in either of the atomic terms A, B,

A7. $x_\alpha \equiv x_\alpha$,

AS8. $B_\alpha \equiv C_\alpha \rightarrow [\,A(B) \rightarrow A(C)\,]$, where $A(B)$, $A(C)$ come from the formula $A(x_\alpha)$ by replacing all free occurrences of x by the atomic terms B , C respectively, and (i) B and C are free for x in $A(x)$, and either (ii) no free occurrence of x in $A(x)$ lies within the scope of $\Box$, or else (ii') B and C are both modally closed,

AS9. $\Box A \rightarrow A$,

AS10. $\Box [A \rightarrow B] \rightarrow [\,\Box A \rightarrow \Box B\,]$,

AS11. $A \rightarrow \Box A$, if A is modally closed,

A12. $\Box [\,{}^{\vee}f_{s\alpha} \equiv {}^{\vee}g_{s\alpha}\,] \rightarrow f \equiv g$.

Rules of Inference.

R1.   From $[A \rightarrow B]$ and A to infer B ,

R2.   From A to infer $\forall x_\alpha A$,

R3.   From A to infer $\Box A$.

We state without proof the following

THEOREM 5.1 (Generalized Completeness Theorem for ML_T)

(i) $\models_g A$ in ML_T if and only if $\vdash A$ in ML_T,

(ii) $\Gamma \models_g A$ in ML_T if and only if $\Gamma \vdash A$ in ML_T,

(iii) Σ is consistent in ML_T if and only if Σ is g-satisfiable in ML_T.

§6. Extensions of IL and ML_T

Since an arbitrary frame and meaning function determine a general model of ML_T, it is natural to seek a set of conditions, formulated in the language of ML_T, which will be satisfied in exactly those g-models of ML_T which are also g-models of IL. For this purpose the theory IL proves to be slightly too weak, however, so we now consider a natural axiomatic extension of it.

Intensional Logic with Descriptors. In both IL and ML_T we introduce the abbreviations

$[A \leftrightarrow B]$ for $[A \to B] \wedge [B \to A]$,

$\exists! x_\alpha\, A$ for $\exists x'_\alpha \forall x_\alpha\, [\, A \leftrightarrow x \equiv x'\,]$,

where x' is the first variable of type α distinct from x and not occurring free in the formula A. We denote by D^e the formula

$\exists f_{(et)e} \forall g_{et}\, [\, \exists! x_e\, [gx] \to g[fg]\,]$,

which we call the <u>axiom of description for individuals</u>. Since D^e is both closed and modally closed, it will be either true or false in any g-model of IL, independent of the index and assignment.

We denote by IL+D the theory obtained from IL by adding D^e as a new axiom, and we write $\vdash A$ in IL+D when the formula A is provable in this theory. By a <u>general model (g-model) of</u> IL+D we understand a g-model of IL in which D^e is true. From Theorem 3.3 it is easy to see that generalized completeness extends to the logic IL+D .

That IL+D is a natural extension of IL is evidenced by the fact[1] that many familiar validities of type-theoretic predicate logic depend for their proof on the axiom D^e . The intuitive content of D^e is just the assertion that there exists a function on sets of individuals, whose value for any singleton is its unique member. Thus, D^e is valid in IL, as are the additional description principles below:

D^α: $\exists f_{(\alpha t)\alpha} \forall g_{\alpha t}\, [\, \exists! x_\alpha\, [gx] \to g[fg]\,]$,

$\hat{D}^\alpha$: $\exists f_{(s(\alpha t))(s\alpha)} \forall g_{s(\alpha t)}\, [\, \Box\, \exists! x_\alpha\, [\check{g}x] \to \Box\, [\check{g}\check{}[fg]]\,]$.

The formula D^α is the analogous principle of description for objects of type α, while $\hat{D}^\alpha$ is an intensional principle of description for such objects. In particular, $\hat{D}^e$ asserts the existence of a function F from properties of individuals to individual concepts, such that for any property G, if it is necessarily true that G is satisfied by exactly one object, then $F(G)$ is the concept of the unique individual satisfying G .

---

[1] Observed in Henkin [1963], p. 343.

The following result, which generalizes a similar result for type theory due to Church, shows that we need not add the formulas D^α, $\hat{D}^\alpha$ to IL+D as axioms.

LEMMA 6.1. For every type $\alpha \in T$, the formulas D^α and $\hat{D}^\alpha$ are provable in IL+D.

Proof: We first use generalized completeness to show that the formula $[D^\alpha \to \hat{D}^\alpha]$ is provable in IL for each α. Let $M = (M_\alpha, m)_{\alpha \in T}$ be a g-model of IL such that M sat D^α. By a rewrite of bound variables (for notational convenience only), we have

$$M \text{ sat } \exists f'_{(at)a} \forall g'_{at} [\exists! x_\alpha [g'x] \to g'[f'g']],$$

and therefore we have

$$M; F' \text{ sat } \forall g'_{at} [\exists! x_\alpha [g'x] \to g'[f'g']]$$

for some $F' \in M_{(at)a}$. No index i need be specified, since the formula in question is modally closed. We now let

$$F = V^M_{F'}(\lambda g_{s(at)} \hat{}[f'_{(at)a}\check{}g]) \in M_{(s(at))(sa)}.$$

It is readily verified that

$$M; F \text{ sat } \forall g_{s(at)} [\Box \exists! x_\alpha [\check{g}x] \to \Box [\check{g}\check{}[fg]]],$$

where f is of type $(s(at))(sa)$, and therefore M sat $\hat{D}^\alpha$. This proves the assertion.

It therefore remains only to show that D^α is provable in IL+D for every α. We use the generalized completeness of IL+D: Suppose $M = (M_\alpha, m)_{\alpha \in T}$ is a g-model of IL+D; we show that M sat D^α by induction on the type α. The case $\alpha = e$ is immediate. For $\alpha = t$ we let

$$F = V^M(\lambda g_{tt} [g \equiv \lambda x_t x_t]) \in M_{(tt)t},$$

and verify that

$$M; F \text{ sat } \forall g_{tt} [\exists! x_t [gx] \to g[f_{(tt)t} g]],$$

from which it follows that M sat D^t (Cf. Henkin [1963], p. 328).

Now assume that M sat D^β; we show that M sat $D^{\alpha\beta}$. First, by a rewrite of bound variables,

$$M \text{ sat } \exists f'_{(\beta t)\beta} \forall g'_{\beta t} [\exists! x_\beta [g'x] \rightarrow g'[f'g']],$$

so that

$$M; F' \text{ sat } \forall g'_{\beta t} [\exists! x_\beta [g'x] \rightarrow g'[f'g']]$$

for some $F' \in M_{(\beta t)\beta}$. Now let $A(f')$ be the term

$$\lambda g_{(\alpha\beta)t} \lambda x_\alpha [f'_{(\beta t)\beta} \lambda y_\beta \exists h_{(\alpha\beta)} [gh \wedge y \equiv hx]]$$

of type $((\alpha\beta)t)(\alpha\beta)$, and let

$$F = V^M_{F'}(A(f')) \in M_{((\alpha\beta)t)(\alpha\beta)}.$$

It is routine to verify that

$$M; F \text{ sat } \forall g_{(\alpha\beta)t} [\exists! h_{\alpha\beta} [gh] \rightarrow g[fg]],$$

where f is of type $((\alpha\beta)t)(\alpha\beta)$, and hence that M sat $D^{\alpha\beta}$ (Cf. Church [1940], p. 62).

Finally, assume that M sat D^α; we show that M sat $D^{s\alpha}$. By a rewrite of bound variables, we have

$$M \text{ sat } \exists f'_{(\alpha t)\alpha} \forall g'_{\alpha t} [\exists! x_\alpha [g'x] \rightarrow g'[f'g']],$$

and hence

$$M; F' \text{ sat } \forall g'_{\alpha t} [\exists! x_\alpha [g'x] \rightarrow g'[f'g']]$$

for some $F' \in M_{(\alpha t)\alpha}$. Let $A(f')$ be the term

$$\lambda g_{(s\alpha)t} \hat{}[f'_{(\alpha t)\alpha} \lambda y_\alpha \exists h_{s\alpha} [gh \wedge y \equiv {\vee}h]]$$

of type $((s\alpha)t)(s\alpha)$, and let

$$F = V^M_{F'}(A(f')) \in M_{((s\alpha)t)(s\alpha)}.$$

One verifies easily that

$$M; F \text{ sat } \forall g_{(s\alpha)t} [\exists! h_{s\alpha} [gh] \rightarrow g[fg]],$$

where f is of type $((s\alpha)t)(s\alpha)$, so that M sat $D^{s\alpha}$. This proves Lemma 6.1.

<u>Modal T-Logic with Replacement</u>. In order to characterize the general models of IL+D it is necessary to add to the theory ML_T certain natural principles of comprehension for functions, which we call <u>replacement principles</u> since they bear a formal resemblance to the replacement schema of axiomatic set theory. The schemata in question are the following:

$R^{\alpha,\beta,A}$: $\forall x_\alpha \exists! y_\beta A \to \exists f_{\alpha\beta} \forall x_\alpha \forall y_\beta [y \equiv fx \to A]$, where f is the first variable of type $\alpha\beta$ not occurring free in the formula A,

$\hat{R}^{\alpha,A}$: $\Box \exists! x_\alpha A \to \exists f_{s\alpha} \Box \forall x_\alpha [x \equiv \check{f} \to A]$, where f is the first variable of type $s\alpha$ not occurring free in the formula A.

We denote by ML_T+R the theory obtained from ML_T by adding all instances of the replacement schemata as new axioms, i.e., by adding all formulas $R^{\alpha,\beta,A}$ and $\hat{R}^{\alpha,A}$ where A is any formula of ML_T. A <u>general model (g-model) of</u> ML_T+R is a g-model of ML_T in which all instances of the replacement schemata are true, i.e., satisfied by every index and assignment. As earlier, it follows from Theorem 5.1 that generalized completeness extends to the present logic, so that, in particular, a formula A of ML_T is provable in ML_T+R if and only if it is g-valid in ML_T+R, i.e., true in every g-model of ML_T+R.

LEMMA 6.2.1. For all types α, $\beta \in T$ and every formula A of IL, the formulas $R^{\alpha,\beta,A}$ and $\hat{R}^{\alpha,A}$ are provable in IL+D.

<u>Proof</u>: We use generalized completeness. Let $M = (M_\alpha, m)_{\alpha \in T}$ be a g-model of IL+D, and let $i \in I$, $a \in As(M)$. We show that M, i, a sat $R^{\alpha,\beta,A}$. Assume that M, i, a sat $\forall x_\alpha \exists! y_\beta A$. Then for every $X \in M_\alpha$ there exists a unique $Y \in M_\beta$ for which $M; i; a,X,Y$ sat A, where a,X,Y abbreviates the assignment $a(x/X)(y/Y)$. By Lemma 6.1, M sat D^β, and therefore

$M; F'$ sat $\forall g_{\beta t} [\exists! y_\beta [gy] \to g[f'_{(\beta t)\beta} g]]$

for some $F' \in M_{(\beta t)\beta}$, where we can assume that the variable f' is distinct from x_α and does not occur free in A. Letting

$$F = V^M_{i;a,F'}(\lambda x_\alpha [f' \lambda y_\beta A]) \in M_{\alpha\beta} ,$$

one easily verifies that

$$M; i; a, F \text{ sat } \forall x_\alpha \forall y_\beta [y \equiv fx \rightarrow A] ,$$

where f is the first variable of type $\alpha\beta$ not occurring free in A , and therefore

$$M, i, a \text{ sat } \exists f \forall x \forall y [y \equiv fx \rightarrow A] ,$$

so that M, i, a sat $R^{\alpha,\beta,A}$.

We show that M, a sat $\hat{R}^{\alpha,A}$: Assume that M, a sat $\Box \exists ! x_\alpha A$, so that for every $i \in I$ there exists a unique $X \in M_\alpha$ such that M; i; a,X sat A . By Lemma 6.1, M sat $\hat{D}^\alpha$, and therefore

$$M; F' \text{ sat } \forall g_{s(\alpha t)} [\Box \exists ! x_\alpha [\check{} gx] \rightarrow \Box [\check{} g\check{} [f'g]]]$$

for some $F' \in M_{(s(\alpha t))(s\alpha)}$, where f' is a variable of type $(s(\alpha t))(s\alpha)$ not occurring free in the formula A . Letting

$$F = V^M_{a,F'}(f' \hat{} \lambda x_\alpha A) \in M_{s\alpha} ,$$

it is straightforward to check that

$$M; a, F \text{ sat } \Box \forall x_\alpha [x \equiv \check{} f \rightarrow A] ,$$

where f is the first variable of type $s\alpha$ not occurring free in A . Therefore

$$M, a \text{ sat } \exists f \Box \forall x [x \equiv \check{} f \rightarrow A] ,$$

which completes the proof of Lemma 6.2.1.

The semantical argument given here can, of course, be effectively replaced by a direct syntactical proof in the theory IL+D , for any given α , β , and A . The lemma shows that, in particular, all instances $R^{\alpha,\beta,A}$ and $\hat{R}^{\alpha,A}$ will be provable in IL+D when A is a formula of the sublanguage ML_T of IL, so by the generalized completeness of IL+D we obtain:

COROLLARY 6.2.2. Every g-model of IL+D is a g-model of ML_T+R .

Before proving the converse to Corollary 6.2.2, which is included in Theorem 6.2, we need an analogue of Lemma 6.2.1. We first observe that the description principles D^α, $\hat{D}^\alpha$ are all formulas of ML_T.

LEMMA 6.2.3. For every type $\alpha \in T$, the formulas D^α and $\hat{D}^\alpha$ are provable in $ML_T + R$.

Proof: Let $M = (M_\alpha, m)_{\alpha \in T}$ be a g-model of $ML_T + R$; we first show that M sat D^α. Let $A(g_{\alpha t}, x_\alpha, x'_\alpha)$ be the formula

$$[\exists! x_\alpha [gx] \land gx] \lor [\sim \exists! x_\alpha [gx] \land x \equiv x']$$

of ML_T. Choosing $X' \in M_\alpha$ arbitrarily, we see that

$$M; X' \text{ sat } \forall g_{\alpha t} \exists! x_\alpha A(g,x,x') .$$

Using replacement and rewrite of bound variables, it follows that

$$M; F,X' \text{ sat } \forall g_{\alpha t} \forall x_\alpha [x \equiv fg \to A(g,x,x')]$$

for some $F \in M_{(\alpha t)\alpha}$, where the variable $f_{(\alpha t)\alpha}$ does not occur free in $A(g,x,x')$. But then clearly

$$M; F \text{ sat } \forall g_{\alpha t} [\exists! x_\alpha [gx] \to g[f_{(\alpha t)\alpha} g]] ,$$

and hence M sat D^α. To see that M sat $\hat{D}^\alpha$, let $A(g_{s(\alpha t)}, h_{s\alpha}, h'_{s\alpha})$ be the formula

$$[\Box \exists! x_\alpha [\check{g}x] \land \Box [\check{g}\check{h}]] \lor [\sim \Box \exists! x_\alpha [\check{g}x] \land h \equiv h']$$

of ML_T. Let $H' \in M_{s\alpha}$ be arbitrary, and suppose that we can show the following condition:

(*) $M; H' \text{ sat } \forall g \exists! h A(g,h,h') .$

Then by replacement and rewrite of bound variables,

$$M; H' \text{ sat } \exists f \forall g \forall h [h \equiv fg \to A(g,h,h')] ,$$

where f is a variable of type $(s(\alpha t))(s\alpha)$ not occurring free in the formula $A(g,h,h')$. We therefore have

$$M; F,H' \text{ sat } \forall g \forall h [h \equiv fg \to A(g,h,h')]$$

for some $F \in M_{(s(\alpha t))(s\alpha)}$; but from this it is straightforward to check that

$$M; F \text{ sat } \forall g_{s(\alpha t)} [\Box \exists! x_\alpha [\check{g}x] \to \Box [\check{g}\check{}[fg]]],$$

where f is of type $(s(\alpha t))(s\alpha)$, and hence that M sat $\hat{D}^\alpha$. So it remains only to prove condition (*). Assume that $G \in M_{s(\alpha t)}$; we show that $M; G, H'$ sat $\exists! h\, A(g,h,h')$. First, in the case $M; G$ sat $\Box \exists! x_\alpha [\check{g}x]$, we have by replacement

$$M; G \text{ sat } \exists h_{s\alpha} \Box \forall x_\alpha [x \equiv \check{}h \to \check{}gx],$$

from which it easily follows that $M; G, H'$ sat $\exists! h\, A(g,h,h')$. On the other hand, this conclusion is immediate when $M; G$ sat $\sim \Box \exists! x_\alpha [\check{g}x]$. This completes the proof of Lemma 6.2.3.

We can now state the main result of this section:

THEOREM 6.2. The logics IL+D and ML_T+R have exactly the same general models.

Proof: In view of Corollary 6.2.2 and Lemma 6.2.3, it suffices to prove the following: Given any g-model $M = (M_\alpha, m)_{\alpha \in T}$ of ML_T+R, there exists in M a value function V^M satisfying the recursive conditions (1) through (7) on page 13.

We first define, for each term A_α of IL and each variable x_α not occurring free in A, a formula

$$Eq^A(x) \quad (x \text{ \underline{equals} } A)$$

of ML_T, whose free variables are x together with the free variables of A_α. The definition is by recursion on A_α:

(i) A_α is v_α, x_α distinct from v_α. Then $Eq^A(x)$ is $[v \equiv x]$.

(ii) A_α is c_α, x_α arbitrary. Then $Eq^A(x)$ is $[c \equiv x]$.

(iii) A_α is $[B_{\beta\alpha} C_\beta]$, y_α not free in A_α. Let $f_{\beta\alpha}$, x_β be the first variables of their respective types which are distinct from y and not free in A, and let $Eq^A(y)$ be the formula

$$\exists f \, \exists x \, [\, Eq^B(f) \wedge Eq^C(x) \wedge y \equiv fx \,] \, .$$

(iv) A_α is $\lambda x_\beta \, B_\gamma$, f_α not free in A_α. Let y_γ be the first variable of type γ which is distinct from x and does not occur free in B, and let $Eq^A(f)$ be the formula

$$\forall x \, \exists y \, [\, Eq^B(y) \wedge fx \equiv y \,] \, .$$

(v) A_α is $[B_\beta \equiv C_\beta]$, x_α not free in A_α. Let y_β, z_β be the first distinct variables of type β which are distinct from x and do not occur free in A, and let $Eq^A(x)$ be the formula

$$\exists y \, \exists z \, [\, Eq^B(y) \wedge Eq^C(z) \wedge [\, x \leftrightarrow y \equiv z \,] \,] \, .$$

(vi) A_α is $^\wedge B_\beta$, f_α not free in A_α. Let x_β be the first variable of type β which does not occur free in B; and let $Eq^A(f)$ be the formula

$$\square \, \exists x \, [\, Eq^B(x) \wedge x \equiv \, ^\vee f \,] \, .$$

(vii) A_α is $^\vee B_{s\alpha}$, x_α not free in A_α. Let $f_{s\alpha}$ be the first variable of type $s\alpha$ which is not free in B, and let $Eq^A(x)$ be the formula

$$\exists f \, [\, Eq^B(f) \wedge x \equiv \, ^\vee f \,] \, .$$

LEMMA 6.2.4. For every term A_α of IL and every variable x_α not occurring free in A, the formula $\exists ! x_\alpha \, Eq^A(x)$ is provable in $ML_T + R$.

The proof is straightforward, using generalized completeness and induction on the term A_α.

Now suppose that $M = (M_\alpha, m)_{\alpha \in T}$ is a g-model of $ML_T + R$. We define a value function V^M in M as follows: Given a term A_α, an index i and an assignment a over M, we let $V^M_{i,a}(A_\alpha)$ be the unique $X \in M_\alpha$ such that $M; i; a, X$ sat $Eq^A(x)$, where x is the first variable of type α not occurring free in A. It is routine to verify that V^M is a value function in M, i.e., satisfies the recursive clauses (1) through (7) of page 13. We omit the details.

Intensional Logic with the Axiom of Choice. We can strengthen the description principles D^α, $\hat{D}^\alpha$ of IL+D by replacing the quantifier $\exists!$ in their antecedents by the weaker existential quantifier $\exists$, obtaining thereby the following axioms of choice:

$Ac^\alpha: \quad \exists f_{(\alpha t)\alpha} \; \forall g_{\alpha t} \; [\; \exists x_\alpha \; [gx] \; \to \; g[fg] \;] \; ,$

$\hat{Ac}^\alpha: \quad \exists f_{(s(\alpha t))(s\alpha)} \; \forall g_{s(\alpha t)} \; [\; \Box \, \exists x_\alpha \; [\check{\;}gx] \; \to \; \Box \; [\check{\;}g\check{\;}[fg]] \;] \; .$

These principles are valid in IL, and if we add them to IL as new axioms we obtain a theory which we denote by IL+Ac. In fact, it suffices to add only the formulas Ac^α, since the intensional axioms $\hat{Ac}^\alpha$ can be shown to follow in IL, as in the proof of Lemma 6.1; however, it does not suffice here to add the formula Ac^e alone. By a general model (g-model) of IL+Ac we understand a g-model of IL in which these axioms of choice are all true. As before, generalized completeness extends to this logic, and clearly the description principles D^α, $\hat{D}^\alpha$ all hold in IL+Ac.

Modal T-Logic with Replacement and Choice. In a similar way we can strengthen our earlier replacement principles $R^{\alpha,\beta,A}$ and $\hat{R}^{\alpha,A}$ to give the following principles of replacement and choice:

$Rc^{\alpha,\beta,A} : \quad \forall x_\alpha \; \exists y_\beta \; A \; \to \; \exists f_{\alpha\beta} \; \forall x \; \forall y \; [\; y \equiv fx \; \to \; A \;] \; ,$

$\hat{Rc}^{\alpha,A} : \quad \Box \, \exists x_\alpha \, A \; \to \; \exists f_{s\alpha} \, \Box \, \forall x_\alpha \; [\; x \equiv \check{\;}f \; \to \; A \;] \; ,$

where in each case f is the first variable of indicated type which does not occur free in the formula A. The theory ML_T+Rc comes from ML_T by adding all instances of these schemata in ML_T to the axioms of ML_T, and a general model (g-model) of ML_T+Rc is defined in the obvious way. By inspecting the proofs of Lemmas 6.2.1, 6.2.3 and Theorem 6.2 one can prove:

THEOREM 6.3. The logics IL+Ac and ML_T+Rc have exactly the same general models.

§7. Normal Forms

The ideas of the previous section can be used to obtain various normal forms for formulas of IL.

THEOREM 7.1. For every formula A of IL we can effectively find a formula A' of ML_T with the same constants and free variables, such that $[A \equiv A']$ is provable in IL.

Proof: For a term A_α of IL and a variable x_α not free in A_α, let $Eq^A(x)$ be the formula of ML_T defined in the proof of Theorem 6.2. One easily shows by induction on A_α:

LEMMA 7.1.1. If x_α is not free in A_α then the formula

$$Eq^A(x) \equiv [A \equiv x]$$

is provable in IL.

Now suppose A is a formula of IL and let x be the first variable of type t which does not occur free in A. Then we can prove in IL the formula

$$A \equiv \exists x \, [\, [A \equiv x] \wedge x \,] \,,$$

so by Lemma 7.1.1 we can also prove

$$A \equiv \exists x \, [\, Eq^A(x) \wedge x \,] \,,$$

and the right-hand side of this equality is the desired formula A'.

COROLLARY 7.2. Let A be a formula of IL, and let A' be the corresponding formula of ML_T, as above. Then:

(i) $\vdash A$ in IL+D if and only if $\vdash A'$ in ML_T+R ,

(ii) $\vdash A$ in IL+Ac if and only if $\vdash A'$ in ML_T+Rc .

Proof: By Theorems 6.2, 7.1 and generalized completeness.

A formula A of ML_T is a <u>prenex</u> formula if it consists of a string of quantifiers followed by a quantifier-free matrix; i.e., A has the form $Q_0 x^0 Q_1 x^1 \ldots Q_{n-1} x^{n-1} M$, where each Q_k is $\forall$ or $\exists$, and the formula M contains no quantifiers. A is a <u>Skolem</u> formula if in addition no universal quantifier precedes an existential quantifier in the prefix, so that A has the form $\exists x^0 \exists x^1 \ldots \exists x^{m-1} \forall x^m \ldots \forall x^{n-1} M$, where the formula M is quantifier-free.

THEOREM 7.3. For every formula A of ML_T we can effectively find a Skolem formula A^* with the same constants and free variables, such that $[A \leftrightarrow A^*]$ is provable in ML_T+Rc .

Proof: Given a prenex formula B of ML_T, we say that B is a <u>prenex form of</u> A if $[A \leftrightarrow B]$ is provable in ML_T+Rc and A and B have the same constants and free variables. We observe that the usual principle of interchange of equivalents holds for the logic ML_T+Rc .

LEMMA 7.3.1. All instances of the schemata

(i) $\forall x_\alpha \exists y_\beta A \leftrightarrow \exists f_{\alpha\beta} \forall x \forall y [y \equiv fx \rightarrow A]$,
(ii) $\Box \exists x_\alpha A \leftrightarrow \exists f_{s\alpha} \Box \forall x_\alpha [x \equiv \check{}f \rightarrow A]$,
(iii) $\Box \forall x_\alpha A \leftrightarrow \forall x_\alpha \Box A$

are provable in ML_T+Rc , where in (i) and (ii) f is a variable which does not occur free in A .

Proof: For (i) and (ii) we only need to establish the converses of the principles of replacement and choice; but these are immediate. (iii) is the so-called <u>Barcan formula</u>, which is easily proved using generalized completeness.

LEMMA 7.3.2. If A and B are prenex formulas of ML_T then for each of the formulas $\sim A$, $[A \wedge B]$, $[A \rightarrow B]$, and $[A \vee B]$ we can effectively find a prenex form.

Proof: As usual.

LEMMA 7.3.3. If A is a prenex formula of ML_T then we can effectively find a prenex form of $\Box A$.

Proof: By induction on the number of quantifiers in the prefix of A . Suppose A is $Q_0 x^0 \ldots Q_{n-1} x^{n-1} M$. If $n = 0$ then $\Box A$ is quantifier-free and hence in prenex form. Otherwise we can clearly assume that the variables $x^0, \ldots, x^{n-1}$ are all distinct, and we have two cases:

Case 1. Q_0 is $\forall$. Then in ML_T+Rc we can prove

(1) $\Box A \leftrightarrow \Box \forall x^0 B$,

(2) $\Box A \leftrightarrow \forall x^0 \Box B$,

by Lemma 7.3.1 (iii), where B is the formula $Q_1 x^1 \ldots Q_{n-1} x^{n-1} M$. But B is a prenex formula with fewer quantifiers than A , so by the induction hypothesis $\Box B$ has a prenex form C , and by (2) the formula $\forall x^0 C$ will be the desired prenex form for $\Box A$.

Case 2. Q_0 is $\exists$. Suppose x^0 is of type α , and write B for the formula $Q_1 x^1 \ldots Q_{n-1} x^{n-1} M$. By Lemma 7.3.1 (ii), (iii) and rewrite of bound variables, we can prove in $ML_T + Rc$:

(1) $\Box A \leftrightarrow \Box \exists x^0_\alpha B$,
(2) $\Box A \leftrightarrow \exists f_{s\alpha} \Box \forall x^0 [x^0 \equiv {}^{\vee}f \to B]$,
(3) $\Box A \leftrightarrow \exists f \forall x^0 \Box [x^0 \equiv {}^{\vee}f \to B]$,

where f is the first variable of type $s\alpha$ which does not occur free in B . Writing $M(x^0, x^1, \ldots, x^{n-1})$ for the matrix M , we can choose new variables $y^1, \ldots, y^{n-1}$, different from x^0 and f , so that in $ML_T + Rc$ we can prove

(4) $\Box A \leftrightarrow \exists f \forall x^0 \Box [x^0 \equiv {}^{\vee}f \to$
           $Q_1 y^1 \ldots Q_{n-1} y^{n-1} M(x^0, y^1, \ldots, y^{n-1})]$,
(5) $\Box A \leftrightarrow \exists f \forall x^0 \Box Q_1 y^1 \ldots Q_{n-1} y^{n-1} [x^0 \equiv {}^{\vee}f \to$
           $M(x^0, y^1, \ldots, y^{n-1})]$.

By the induction hypothesis, therefore, we can find a prenex formula C such that the formula

$\Box A \leftrightarrow \exists f \forall x^0 C$

is provable in $ML_T + Rc$, and this gives the desired prenex form for $\Box A$.

By Lemmas 7.3.2, 7.3.3 and a straightforward induction on A we have:

LEMMA 7.3.4. For every formula A of ML_T we can effectively find a prenex form.

To prove Theorem 7.3 it clearly suffices to combine Lemma 7.3.4 with the following result:

LEMMA 7.3.5. Let A be a Skolem formula with n existential quantifiers in its prefix. Then we can effectively find a Skolem form of $\forall x_\alpha\, A$ having at most n existential quantifiers in its prefix.

Proof: By induction on the number n. Suppose A is of the form $\exists y^0 \ldots \exists y^{n-1}\, B$, where B is the formula $\forall z^0 \ldots \forall z^{m-1}\, M$. Clearly we can assume $n > 0$ and $x, y^0, \ldots, z^{m-1}$ distinct, by dropping any vacuous quantifiers. Using Lemma 7.3.1 (i) and rewrite of bound variables, we can prove in $ML_T + Rc$ the following formulas:

(1) $\quad \forall x_\alpha\, A \leftrightarrow \forall x_\alpha\, \exists y^0_\beta\, \exists y^1 \ldots \exists y^{n-1}\, B$,

(2) $\quad \forall x_\alpha\, A \leftrightarrow \exists f_{\alpha\beta}\, \forall x\, \forall y^0\, [\, y^0 \equiv fx \to \exists y^1 \ldots \exists y^{n-1}\, B\,]$,

where f is the first variable of type $\alpha\beta$ not occurring free in B. Writing $M(x, y^0, \ldots, z^0, \ldots)$ for M, we can choose new variables $u^1, \ldots, u^{n-1}, v^0, \ldots, v^{m-1}$ different from x, y^0 and f, so that in $ML_T + Rc$ we can prove

$$\forall x\, A \leftrightarrow \exists f\, \forall x\, \forall y^0\, C,$$

where C is the formula

$$\exists u^1 \ldots \exists u^{n-1}\, \forall v^0 \ldots \forall v^{m-1}\, [\, y^0 \equiv fx \to M(x, y^0, u^1, \ldots, v^0, \ldots)\,].$$

Since C is a Skolem formula with $n-1$ existential quantifiers, two applications of the induction hypothesis give a Skolem formula C' with at most $n-1$ existential quantifiers in its prefix, such that

$$\forall x\, A \leftrightarrow \exists f\, C'$$

is provable in $ML_T + Rc$. Thus $\exists f\, C'$ is the desired Skolem form of $\forall x\, A$.

COROLLARY 7.4. For every formula A of IL we can effectively find a Skolem formula A^* of ML_T with the same constants and free variables, such that $[A \equiv A^*]$ is provable in $IL + Ac$.

Proof: Theorems 6.3, 7.1 and 7.3.

REMARK: Dual to this existential Skolem form we have a universal Skolem form, in which no existential quantifier precedes a universal quantifier. The corresponding theorems follow from Theorem 7.3 and Corollary 7.4

by considering the existential Skolem form of $\sim A$. It should be noted also that the matrix of a prenex formula can be put in various modal normal forms[1] on the basis of the S5 axioms of ML_T.

§8. Two-Sorted Type Theory

As we observed in §2, the cap operator $\char`^$ acts as a functional abstractor over indices, although the grammar of IL lacks variables over indices since s alone is not a type. This omission is reasonable, since IL was intended as a formal logic with intensional features close to those of natural language, and in natural language we do not refer explicitly to contexts of use; indeed, if we did refer to them explicitly there would be little justification for the Carnap approach. From a formal point of view, however, it is natural to consider interpreting IL in an extensional theory of types having two sorts of individuals. We call this logic <u>Two-Sorted Type Theory</u>, and denote it by Ty_2.

<u>Types</u>. The set T_2 of <u>types of</u> Ty_2 is the smallest set such that:

(i) $e, t, s \in T_2$,

(ii) $\alpha, \beta \in T_2$ imply $(\alpha, \beta) \in T_2$.

Thus, the set T of types of IL is contained in the set T_2.

<u>Primitive Symbols</u>. For each $\alpha \in T_2$, we admit <u>variables</u>

$$x_\alpha^0, x_\alpha^1, x_\alpha^2, \ldots$$

and non-logical <u>constants</u>

$$c_\alpha^0, c_\alpha^1, c_\alpha^2, \ldots$$

of type α, which we identify with the corresponding symbols of IL when the type α belongs to T. We also have the improper symbols $\equiv$, λ, [and]. As before, we denote the first nine variables of type α by:

---

[1] E.g., the modal conjunctive normal form described in Hughes and Cresswell [1968], pp. 54-56.

x_α, y_α, z_α, u_α, v_α, w_α, f_α, g_α, h_α.

Terms. The sets $Tm_{2,\alpha}$ of terms of Ty_2 of type α are characterized recursively:[1]

(i) Variables and constants of type α belong to $Tm_{2,\alpha}$,

(ii) $A \in Tm_{2,\alpha\beta}$, $B \in Tm_{2,\alpha}$ imply $[AB] \in Tm_{2,\beta}$,

(iii) $A \in Tm_{2,\beta}$ implies $\lambda x_\alpha A \in Tm_{2,\alpha\beta}$,

(iv) A, $B \in Tm_{2,\alpha}$ imply $[A \equiv B] \in Tm_{2,t}$.

Generalized Semantics. Let D and I be non-empty sets. By a frame for Ty_2 based on D and I we understand an indexed family $(M_\alpha)_{\alpha \in T_2}$ of sets, where

(i) $M_e = D$,

(ii) $M_t = 2 = \{0,1\}$,

(iii) $M_s = I$,

(iv) $M_{\alpha\beta}$ is a non-empty subset of $M_\beta^{M_\alpha}$.

The frame is standard if the inclusion in condition (iv) can be replaced by equality. A general model (g-model) of Ty_2 based on D and I is a system $M = (M_\alpha, m)_{\alpha \in T_2}$ satisfying:

(i) $(M_\alpha)_{\alpha \in T_2}$ is a frame for Ty_2 based on D and I,

(ii) $m(c_\alpha) \in M_\alpha$ for each constant c_α,

(iii) There exists a function V^M which assigns, to each assignment a over M and each term A_α, a value $V_a^M(A_\alpha) \in M_\alpha$, in such a way that the following conditions hold:

(1) $V_a(x_\alpha) = a(x_\alpha)$,

(2) $V_a(c_\alpha) = m(c_\alpha)$,

---

[1] We employ freely in this section various of our notational conventions for the logic IL.

(3) $\quad V_a(A_{\alpha\beta} B_\alpha) = V_a(A_{\alpha\beta})[V_a(B_\alpha)]$,

(4) $\quad V_a(\lambda x_\alpha A_\beta) =$ the function F on M_α whose value at $X \in M_\alpha$ is equal to $V_{a'}(A_\beta)$, where $a' = a(x/X)$,

(5) $\quad V_a(A_\alpha \equiv B_\alpha) = 1$ if $V_a(A_\alpha) = V_a(B_\alpha)$, and 0 otherwise.

If the underlying frame is standard then condition (iii) is unnecessary, and M is called a (standard) model of Ty_2. As before, a formula is a term A of type t. The notions M, a sat A, $\Gamma \models_g A$ in Ty_2, $\models_g A$ in Ty_2, and Σ is g-satisfiable in Ty_2, are defined in the usual way, as are their standard semantical counterparts, e.g., the notion $\Gamma \models A$ in Ty_2. Also, we employ in Ty_2 the definitions of the logical operators T, F, $\sim$, $\wedge$, $\rightarrow$, $\vee$, $\forall$, $\exists$ given in §2.

The Theory Ty_2.

Axioms of Ty_2.

A1. $\quad g_{tt} T \wedge g_{tt} F \equiv \forall x_t [gx]$,

A2. $\quad x_\alpha \equiv y_\alpha \rightarrow f_{at} x \equiv f_{at} y$,

A3. $\quad \forall x_\alpha [f_{\alpha\beta} x \equiv g_{\alpha\beta} x] \equiv [f \equiv g]$,

AS4. $\quad (\lambda x_\alpha A_\beta(x)) B_\alpha \equiv A_\beta(B_\alpha)$, where $A_\beta(B_\alpha)$ comes from $A_\beta(x_\alpha)$ by replacing all free occurrences of x by the term B, and B is free for x in $A(x)$.

Rule of Inference.

R. From $A_\alpha \equiv A'_\alpha$ and the formula B to infer the formula B', where B' comes from B by replacing one occurrence of A (not immediately preceded by λ) by the term A'.

We have generalized completeness for the two-sorted logic Ty_2, as a trivial extension of Henkin's result for ordinary type theory. It is worth noting, in particular, that the schemata

(i) $\quad \forall x_\alpha A(x) \rightarrow A(B_\alpha)$, where the term B is free for x in the formula $A(x)$,

(ii) $B_\alpha \equiv C_\alpha \to A_\beta(B) \equiv A_\beta(C)$, where B and C are free for x in the term $A_\beta(x_\alpha)$,

are provable in Ty_2 without further restriction (Cf. discussion at the end of §2).

We denote by Ty_2+D the theory obtained from Ty_2 by adding as new axioms the formulas

D^e: $\exists f_{(et)e} \forall g_{et} [\exists! x_e [gx] \to g[fg]]$,

D^s: $\exists f_{(st)s} \forall g_{st} [\exists! x_s [gx] \to g[fg]]$.

As before we can prove:

LEMMA 8.1. In Ty_2+D the formulas

D^α: $\exists f_{(\alpha t)\alpha} \forall g_{\alpha t} [\exists! x_\alpha [gx] \to g[fg]]$

are provable for each type $\alpha \in T_2$.

<u>Interpretability of IL in</u> Ty_2. For each term A_α of IL we define A_α^*, the translate of A_α in Ty_2, as follows:

(i) $[x_\alpha^n]^* = x_\alpha^n$,

(ii) $[c_\alpha^n]^* = [c_{s\alpha}^n x_s]$,

(iii) $[A_{\alpha\beta} B_\alpha]^* = [A^* B^*]$,

(iv) $[\lambda x_\alpha A_\beta]^* = \lambda x A^*$,

(v) $[A_\alpha \equiv B_\alpha]^* = [A^* \equiv B^*]$,

(vi) $[{^\wedge}A_\alpha]^* = \lambda x_s A^*$,

(vii) $[{^\vee}A_{s\alpha}]^* = [A^* x_s]$.

The free variables of A_α^* are just the free variables of A_α together, in some cases, with the single variable x_s . The constants of A^* are the constants $c_{s\beta}^n$ such that c_β^n occurs in A . If Γ is a set of formulas of IL, we denote by Γ^* the set of formulas A^* for $A \in \Gamma$.

THEOREM 8.2. The translation of A into A^* preserves the standard semantics. Precisely, let Γ and Σ be sets of formulas of IL, A a formula of IL. Then

(i) $\models A$ in IL if and only if $\models A^*$ in Ty_2,

(ii) $\Gamma \models A$ in IL if and only if $\Gamma^* \models A^*$ in Ty_2,

(iii) Σ satisfiable in IL if and only if Σ^* satisfiable in Ty_2.

Proof: (i) and (ii) follow from (iii), which in turn follows from the following

LEMMA 8.2.1. Let D and I be non-empty sets, and suppose that $M = (M_\alpha, m)_{\alpha \in T}$, $M^* = (M^*_\alpha, m^*)_{\alpha \in T_2}$ are standard models of IL and Ty_2, respectively, based on D and I, so that $M_\alpha = M^*_\alpha$ for $\alpha \in T$. Suppose also that $m(c^n_\alpha) = m^*(c^n_{s\alpha})$ for each constant c^n_α of IL. Then for every term A_α of IL, every assignment a over M and index $i \in I$:

$$V^M_{i,a}(A_\alpha) = V^{M^*}_{a^*}(A^*_\alpha),$$

where a^* is the partial assignment $a(x_s/i)$ over M^*.

Proof: Straightforward induction on A_α.

Less obvious than Theorem 8.2 is the fact that the translation of A into A^* provides a relative interpretation, in a sense close to that of Tarski, Mostowski and Robinson [1953], of the theory IL+D in the theory Ty_2+D. Precisely:

THEOREM 8.3. Let Γ and Σ be sets of formulas and let A be a formula of IL. Then:

(i) $\vdash A$ in IL+D implies $\vdash A^*$ in Ty_2+D,

(ii) $\Gamma \vdash A$ in IL+D implies $\Gamma^* \vdash A^*$ in Ty_2+D,

(iii) Σ^* consistent in Ty_2+D implies Σ consistent in IL+D.

Proof: Again (i) and (ii) follow from (iii). By generalized completeness it suffices to show:

LEMMA 8.3.1. If Σ^* is g-satisfiable in Ty_2+D, then Σ is g-satisfiable in $IL+D$.

Proof: Let $M^* = (M_\alpha^*, m^*)_{\alpha \in T_2}$ be a g-model of Ty_2+D in which Σ^* is satisfiable, based on sets D and I. Define a g-model $M = (M_\alpha, m)_{\alpha \in T}$ of IL based on D and I by letting $M_\alpha = M_\alpha^*$ for $\alpha \in T$ and putting $m(c_\alpha^n) = m^*(c_{s\alpha}^n)$. For $a \in As(M)$ and $i \in I$ let

$$V_{i,a}^M(A_\alpha) = V_{a^*}^{M^*}(A_\alpha^*),$$

where a^* is the partial assignment $a(x_s/i)$. It is easily checked that V^M is a value function in M, and clearly for every formula A of IL

(†) M, i, a sat A if and only if M^*, a^* sat A^*.

But Σ^* is satisfiable in M^*, and in fact we can assume that M^*, a^* sat Σ^* for some a^* of the form $a(x_s/i)$, where $a \in As(M)$ and $i \in I$. But then by (†), M, i, a sat Σ. It therefore remains only to show that M is a g-model of $IL+D$, i.e., that M sat D^e. But it is clear that $[D^e]^*$ is D^e, and M^* sat D^e since M^* is a g-model of Ty_2+D. By (†), the proof is therefore complete.

We conclude with two remarks. First, it is possible to interpret the theory Ty_2+D in the theory $IL+D$ in a similar sense, using notions to be developed in the next chapter; we shall return to this question briefly in §13. Second, each theory is <u>strongly</u> interpretable in the other, in the sense that the implications in Theorem 8.3, for example, can actually be strengthened to equivalence. We omit the very lengthy proof of this fact, although the general idea is discussed at the end of §13.

PART II. HIGHER-ORDER MODAL LOGIC

CHAPTER 3. HIGHER-ORDER MODAL LOGIC

§9. Modal Predicate Logic

We now consider another alternative formulation of IL, which we call Modal Predicate Logic and denote by ML_p. Like the system ML_T of §5, this logic takes ∀ and □ as primitives; unlike ML_T, however, its types are restricted to include only those for individuals and predicates at various levels. Here predicate is used in a precise sense employed by Montague[1] to mean relation-in-intension. Thus, an n-place predicate is to an n-place relation what a property is to a set. Such a restriction of the set of types seems natural to a formulation in which ∀ and □ are primitive, and it is perhaps not surprising that several authors have proceeded along these lines in generalizing modal predicate logic to various higher orders. Bayart [1959] and Cocchiarella [1969] give generalized completeness theorems for systems of second-order S5; Bayart's methods, however, do not seem to generalize readily to higher orders. Bressan [1964] has applied higher-order S5 to problems arising in the foundations of physics, and in his most recent work [1972] he develops in detail a logic similar to ML_p, allowing unlimited predicate types. Montague [1970a] independently employed a second-order modal logic in connection with his analysis of belief contexts, mentioned in §1, and remarked that the same construction could be carried to higher (and even transfinite) orders. The logic ML_p is therefore a natural and useful alternative to IL; moreover, we shall see that ML_p has some distinct advantages over IL when we come to consider the Boolean semantics of Chapter 4.

Higher-Order Predicate Logic. Before defining the syntax and semantics of ML_p, we consider a formulation of ordinary (non-modal) higher-order predicate logic, which we denote by L_p. This logic, which is essentially the version presented in Orey [1959], will be useful in its own right in

---

[1] Montague [1970a], p. 71.

a later section, and its syntax and semantics will be closely paralleled by those of the logic ML_p.

Predicate Types. Let e be any symbol which is not a finite sequence. The set P of <u>predicate types</u> is the smallest set such that:

(i) $e \in P$,

(ii) $\sigma_0, \sigma_1, \ldots, \sigma_{n-1} \in P$ imply $(\sigma_0, \sigma_1, \ldots, \sigma_{n-1}) \in P$.

That is, the set of predicate types contains e and is closed under the formation of arbitrary finite sequences. Objects of type e will be individuals, and objects of type $(\sigma_0, \sigma_1, \ldots, \sigma_{n-1})$ will be relations of n arguments, of which the first is an object of type σ_0, the second an object of type σ_1, etc.

Primitive Symbols. For each $\sigma \in P$ we have a denumerable list of <u>variables</u>

$$x_\sigma^0, x_\sigma^1, x_\sigma^2, \ldots$$

and non-logical <u>constants</u>[2]

$$c_\sigma^0, c_\sigma^1, c_\sigma^2, \ldots$$

of type σ, together with the improper symbols $\equiv, \sim, \rightarrow, \forall, [,]$. We also denote the variables of type σ, in their proper order, by

$$x_\sigma, y_\sigma, z_\sigma, u_\sigma, v_\sigma, w_\sigma, f_\sigma, g_\sigma, h_\sigma, p_\sigma, q_\sigma, r_\sigma,$$
$$x_\sigma', y_\sigma', z_\sigma', \ldots,$$

and we use the letters 'x', 'y', ... , 'r', with or without superscripts or primes, to range over formal variables of L_p. A <u>symbol</u> s_σ <u>of type</u> σ is a variable or constant of that type.

Grammar. An <u>atomic formula</u> of L_p is an expression of one of the forms

$$s \; s^0 s^1 \ldots s^{n-1},$$

where s is of type $\sigma = (\sigma_0, \sigma_1, \ldots, \sigma_{n-1})$ and s^k is a symbol of type

---

[2] We fix the set of constants here for reasons of convenience. One could allow an arbitrary set of constants, not necessarily denumerable.

MODAL PREDICATE LOGIC   69

σ_k for $k < n$; or

$$[s \equiv s'] ,$$

where s, s' are symbols of type e. The <u>formulas</u> of L_p are generated from the atomic formulas by the connectives $\sim$, $\to$ and the quantifier $\forall x_\sigma$, where x_σ is an arbitrary variable.

It is important to note that the <u>empty sequence</u> ϕ belongs to P, so that a symbol s_ϕ standing alone is an atomic formula.

The sentential connectives $\wedge$, $\vee$, $\leftrightarrow$ and the quantifier $\exists x_\sigma$ are defined as usual. For an arbitrary predicate type $\sigma \neq e$ and symbols s, s' of type σ we use $[s \equiv s']$ as an abbreviation for the formula

$$\forall x^0 \forall x^1 \ldots \forall x^{n-1} [s x^0 x^1 \ldots x^{n-1} \leftrightarrow s' x^0 x^1 \ldots x^{n-1}] ,$$

where $\sigma = (\sigma_0, \sigma_1, \ldots, \sigma_{n-1})$ and x^k is of type σ_k for $k < n$. We use $\exists ! x_\sigma A$ as an abbreviation for the formula

$$\exists x'_\sigma \forall x_\sigma [A \leftrightarrow x \equiv x'] ,$$

where x'_σ is the first variable of type σ different from x and not occurring free in the formula A.

<u>Generalized Semantics</u>. Given a set X, we denote by $P(X)$ the power set, or set of all subsets, of X. Given sets $X_0, \ldots, X_{n-1}$, we let $X_0 \times \ldots \times X_{n-1}$ denote their Cartesian product, i.e., the set of all sequences $(a_0, \ldots, a_{n-1})$, where $a_k \in X_k$ for $k < n$.

Let D be a non-empty set. By a <u>frame for</u> L_p <u>based on</u> D we understand an indexed family $(M_\sigma)_{\sigma \in P}$ of sets, where

(i) $M_e = D$,

(ii) For each type $\sigma = (\sigma_0, \ldots, \sigma_{n-1})$, M_σ is a non-empty subset of $P(M_{\sigma_0} \times \ldots \times M_{\sigma_{n-1}})$.

The frame is <u>standard</u> if the inclusion in (ii) can be replaced by equality. A <u>general model (g-model) of</u> L_p <u>based on</u> D is a system $M = (M_\sigma, m)_{\sigma \in P}$ such that:

(i) $(M_\sigma)_{\sigma \in P}$ is a frame for L_p based on D ,

(ii)   The mapping m assigns to each constant c_σ an element of M_σ .

M is a <u>(standard) model of</u> L_p if the underlying frame is standard. We denote by As(M) the set of all <u>assignments over</u> the g-model M , i.e., all functions a on the set of variables such that $a(x_\sigma) \in M_\sigma$ for each variable x_σ . For an assignment a , we let $\bar{a}$ be the extension of a to the set of all constants, defined by the rule that $\bar{a}(c_\sigma) = m(c_\sigma) \in M_\sigma$. We can define the notion

    M, a sat A

by recursion on the formula A of L_p, as follows:

(i)   M, a sat $s\, s^0 \ldots s^{n-1}$ if and only if $(\bar{a}(s^0), \ldots, \bar{a}(s^{n-1})) \in \bar{a}(s)$,

(ii)   M, a sat $[s \equiv s']$ if and only if $\bar{a}(s) = \bar{a}(s')$, where s and s' are symbols of type e ,

(iii)   Usual satisfaction clauses for $\sim$, $\to$, $\forall x_\sigma$.

It is readily verified that the defined equality relation $[s \equiv s']$ for symbols of type $\sigma \neq e$ represents identity in any g-model of L_p, in the sense that M, a sat $[s \equiv s']$ if and only if $\bar{a}(s) = \bar{a}(s')$. From this it follows that M, a sat $\exists! x_\sigma A$ if and only if there exists a unique $X \in M_\sigma$ for which M; a,X sat A . (As in earlier sections, a,X is here an abbreviation for the assignment a(x/X) .) We define as usual the semantical notions: A is <u>true in</u> M , A is a <u>g-semantical consequence of</u> Γ in L_p, A is <u>g-valid</u> in L_p, etc.

<u>The Theory</u> L_p.

<u>Axioms of</u> L_p.

AS1.   A , where A is tautologous in $\sim$ and $\to$,

AS2. $\forall x_\alpha [A \to B] \to [A \to \forall x\, B]$, where x is any variable not occurring free in the formula A ,

AS3. $\forall x_\sigma A(x) \to A(s_\sigma)$, where the symbol s is free for x in the formula A(x) ,

A4. $x_e \equiv x_e$,

A5. $s_\sigma \equiv s'_\sigma \to [\, A(s) \to A(s')\,]$, where the symbols s and s' are free for x_σ in the formula $A(x_\sigma)$.

Rules of Inference.

R1. From $[A \to B]$ and A to infer B ,

R2. From A to infer $\forall x_\sigma\, A$.

It is well-known[3] that generalized completeness holds for the logic L_p, as does the corresponding result for the logic $L_p + C$, <u>Predicate Logic with Comprehension</u>, obtained by adding to the axioms of L_p all instances, in the language of L_p, of the following schema:

$$C^{\sigma, A} : \quad \exists f_\sigma\, \forall x^0\, \forall x^1 \ldots \forall x^{n-1}\, [\, f\, x^0 x^1 \ldots x^{n-1} \leftrightarrow A\,] \;,$$

where $\sigma = (\sigma_0, \sigma_1, \ldots, \sigma_{n-1})$, x^k is of type σ_k for $k < n$, and f_σ is the first variable of type σ which is not free in the formula A .

<u>Modal Predicate Logic.</u> As indicated earlier, the syntax and semantics of ML_p closely parallel the syntax and semantics of L_p. In fact, the set P of predicate types is the same for the two logics, the difference lying in their intended interpretation. In ML_p, objects of type $(\sigma_0, \sigma_1, \ldots, \sigma_{n-1})$ will be predicates (relations-in-intension) of n arguments, of which the first is an object of type σ_0 , the second an object of type σ_1 , etc.

<u>Grammar.</u> The <u>variables</u> and <u>constants</u> of ML_p are the same as those of L_p. The improper symbols of ML_p are those of L_p together with the necessity operator $\square$. The <u>formulas</u> of ML_p are generated from the atomic formulas given earlier by means of the operators $\sim$, $\to$, $\forall x_\sigma$ and $\square$. The sentential connectives $\wedge$, $\vee$, $\leftrightarrow$, the quantifier $\exists x_\sigma$, and the possibility operator $\diamond$ are defined as usual. We carry over from L_p the abbreviations $[s \equiv s']$ for symbols of type $\sigma \neq e$, and $\exists! x_\sigma A$ (given earlier). In ML_p we also write

$$[s \equiv s'] \quad \text{for} \quad \square\, [s \equiv s'] \;,$$

---

[3] By the method of Henkin [1950].

where s and s' are symbols of arbitrary type σ, and

$$\exists!!x_\sigma A \text{ for } \exists x'_\sigma \forall x_\sigma [A \leftrightarrow x \equiv x'],$$

where x'_σ is the first variable of type σ different from x and not free in the formula A.

<u>Generalized Semantics</u>. Let D and I be non-empty sets. A <u>frame for</u> ML_p <u>based on</u> D <u>and</u> I is an indexed family $(M_\sigma)_{\sigma \in P}$ of sets, where

(i) $M_e = D$,

(ii) For each type $\sigma = (\sigma_0, \ldots, \sigma_{n-1})$, M_σ is a non-empty subset of $P(M_{\sigma_0} \times \ldots \times M_{\sigma_{n-1}})^I$.

The frame is <u>standard</u> if equality holds in (ii). A <u>general model (g-model)</u> <u>of</u> ML_p <u>based on</u> D <u>and</u> I is a system $M = (M_\sigma, m)_{\sigma \in P}$ such that:

(i) $(M_\sigma)_{\sigma \in P}$ is a frame for ML_p based on D and I,

(ii) The mapping m assigns to each constant c_σ an element of M_σ.

If $n = 0$ we adopt the usual set-theoretic convention identifying the Cartesian product $X_0 \times \ldots \times X_{n-1}$ with the set containing only the empty sequence ϕ. In any g-model M of ML_p we therefore have

$$M_\phi \subseteq P(\{\phi\})^I = 2^I,$$

so that M_ϕ is always a non-empty set of propositions. A <u>(standard) model</u> <u>of</u> ML_p is a g-model whose underlying frame is standard. An <u>assignment</u> is defined as before, and the notion

$$M, i, a \text{ sat } A,$$

where $i \in I$ and $a \in As(M)$, is defined by recursion on the formula A:

(i) M, i, a sat $s \, s^0 \ldots s^{n-1}$ if and only if $(\bar{a}(s^0), \ldots, \bar{a}(s^{n-1}))$ is an element of $\bar{a}(s)(i)$,

(ii) M, i, a sat $[s \equiv s']$ if and only if $\bar{a}(s) = \bar{a}(s')$, where s and s' are symbols of type e,

(iii) Usual satisfaction clauses for $\sim$, $\to$, $\forall x_\sigma$,

(iv)   M, i, a sat $\square$ A if and only if M, j, a sat A for all $j \in I$.

The defined equality relation $[s_\sigma \equiv s'_\sigma]$ for types $\sigma \neq e$ now represents <u>contingent</u> identity of predicates in any g-model of ML_p: We have M, i, a sat $[s_\sigma \equiv s'_\sigma]$ if and only if $\bar{a}(s)(i) = \bar{a}(s')(i)$. But for every type σ, M, i, a sat $[s_\sigma \equiv s'_\sigma]$ if and only if $\bar{a}(s) = \bar{a}(s')$. It is also easily checked that M, i, a sat $\exists ! x_\sigma$ A just in case (i) M; i; a,X sat A for some $X \in M_\sigma$, and (ii) the condition M; i; a,X sat A determines $X(i)$ uniquely. On the other hand, M, i, a sat $\exists !! x_\sigma$ A just in case there exists a unique $X \in M_\sigma$ for which M; i; a,X sat A .

As in §2 and §3 we introduce the notions A is <u>true in</u> M , $\Gamma \models_g A$ in ML_p, $\models_g A$ in ML_p, and Σ is <u>g-satisfiable</u> in ML_p. We also have the corresponding standard semantical notions $\Gamma \models A$ in ML_p, $\models A$ in ML_p, and Σ is <u>satisfiable</u> in ML_p. The set of <u>modally closed</u> formulas of ML_p is the smallest set containing all atomic formulas of the form $[s_e \equiv s'_e]$, all formulas of the form $\square$ A , and closed under the connectives $\sim$, $\to$ and the quantifier $\forall x_\sigma$. For such a formula A we write M, a sat A , as earlier, since the index i is irrelevant.

<u>The Theory</u> ML_p.

<u>Axioms of</u> ML_p.

AS1.   A , where A is tautologous in $\sim$ and $\to$,

AS2. $\forall x_\sigma [A \to B] \to [A \to \forall x B]$, where x is any variable not occurring free in the formula A ,

AS3. $\forall x_\sigma A(x) \to A(s_\sigma)$, where the symbol s is free for x in the formula $A(x)$,

A4. $x_e \equiv x_e$,

A5. $x_e \equiv y_e \to x_e \equiv y_e$,

AS6. $s_\sigma \equiv s'_\sigma \to [A(s) \to A(s')]$, where the symbols s and s' are free for x_σ in the formula $A(x_\sigma)$,

AS7. $\square A \to A$,

AS8. $\Box [A \to B] \to [\Box A \to \Box B]$,

AS9. $\sim \Box A \to \Box \sim \Box A$.

Rules of Inference.

R1.   From $[A \to B]$ and A to infer B ,

R2.   From A to infer $\forall x_\sigma A$,

R3.   From A to infer $\Box A$.

We write $\vdash A$ in ML_p, if the formula A is provable in this theory, and $\Gamma \vdash A$ in ML_p, if the formula

$$B^0 \to . B^1 \to . \ldots \to . B^{n-1} \to A$$

is provable in ML_p for some formulas $B^0, B^1, \ldots, B^{n-1}$ in Γ. A set Σ of formulas is <u>consistent</u> in ML_p if some formula is not derivable from Σ in ML_p. The soundness of the theory ML_p relative to the generalized semantics for ML_p is easily established using the following straightforward semantical lemma:

LEMMA 9.1.1. Let M be a g-model of ML_p, and suppose the symbol s_σ is free for the variable x_σ in the formula $A(x)$. Then for every index i and assignment a ,

M, i, a sat $A(s)$ if and only if $M; i; a,X$ sat $A(x)$,

where $X = \bar{a}(s)$.

THEOREM 9.1 (Generalized Completeness Theorem for ML_p)

(i) $\vDash_g A$ in ML_p if and only if $\vdash A$ in ML_p,

(ii) $\Gamma \vDash_g A$ in ML_p if and only if $\Gamma \vdash A$ in ML_p,

(iii) Σ is consistent in ML_p if and only if Σ is g-satisfiable in ML_p.

We sketch briefly the proof, which is considerably simpler than the proof of Theorem 3.3. As earlier, it suffices to prove the implication from left to right in part (iii), and again we can assume that the consistent set Σ omits infinitely many variables of each type σ. Lemma 3.2 carries over to the theory ML_p (see comment on pp. 29-30), so there is a sequence

$\overline{\Sigma} = (\overline{\Sigma}_i)_{i \in \omega}$ of sets of formulas of ML_p having properties (i) through (iv) of Lemma 3.2 (see page 25) and hence also properties (v) and (vi) of Remark 3.2.9 (page 29). Given symbols s , s' of type σ , the relation

$$s \simeq s' \quad \text{if and only if} \quad [s \equiv s'] \in \overline{\Sigma}_i \ ,$$

which is independent of $i \in \omega$, is easily shown to be an equivalence relation on the set Sym_σ of symbols of type σ . Moreover, for each symbol s_σ we have $s_\sigma \simeq x_\sigma$ for infinitely many variables x_σ . By recursion on the type σ we define a set M_σ and a mapping μ_σ from Sym_σ into M_σ such that:

(1) μ_σ is onto M_σ ,

(2) $\mu_\sigma(s_\sigma) = \mu_\sigma(s'_\sigma)$ if and only if $s_\sigma \simeq s'_\sigma$.

We first let $M_e = D = Sym_e/\simeq$ and define $\mu_e(s_e)$ to be $s_e/\simeq$. Next, we assume that M_{σ_k} and μ_{σ_k} have been defined for $k < n$; we define a mapping μ_σ from Sym_σ into $P(M_{\sigma_0} \times \ldots \times M_{\sigma_{n-1}})^\omega$ where $\sigma = (\sigma_0, \ldots, \sigma_{n-1})$ by putting the sequence

$$(\mu_{\sigma_0}(s^0_{\sigma_0}) \ , \ \ldots \ , \ \mu_{\sigma_{n-1}}(s^{n-1}_{\sigma_{n-1}}) \ )$$

into $\mu_\sigma(s_\sigma)(i)$ just in case the formula $s\, s^0 \ldots s^{n-1}$ belongs to $\overline{\Sigma}_i$. This is well-defined, by AS6, and if we let M_σ be the range of μ_σ then conditions (1) and (2) hold. We define a g-model $M = (M_\sigma, m)_{\sigma \in P}$ of ML_p based on D and $I = \omega$ by letting $m(c_\sigma) = \mu_\sigma(c_\sigma)$ for each constant c_σ. It is readily verified by induction on the length of the formula A that

$$M, i, \mu \text{ sat } A \quad \text{if and only if} \quad A \in \overline{\Sigma}_i \ ,$$

for every $i \in I$, using Lemma 9.1.1 and property (v) of $\overline{\Sigma}$ at the quantifier step. From this we conclude that M, i, a sat Σ when $i = 0$ and $a = \mu$, and the proof is complete.

<u>Persistence in</u> ML_p. The notion of persistence, discussed in §4, also carries over to ML_p in a much simpler form. Suppose $M = (M_\sigma, m)_{\sigma \in P}$ is a g-model of ML_p based on D and I , and let $(M'_\sigma)_{\sigma \in P}$ be the standard frame for ML_p based on D and I . It is easily seen that $M_\sigma \subseteq M'_\sigma$ for every $\sigma \in P$, so that the system $M' = (M'_\sigma, m)_{\sigma \in P}$ is a standard model of

ML_p and $As(M) \subseteq As(M')$. A formula A of ML_p is called M-<u>persistent</u> if

$$M, i, a \text{ sat } A \text{ if and only if } M', i, a \text{ sat } A$$

for every $i \in I$ and $a \in As(M)$, and <u>persistent</u> if it is M-persistent for every g-model M of ML_p. Any formula which is provably equivalent to a persistent formula is itself persistent, and as earlier we can prove:

THEOREM 9.2. Let Per be the set of all persistent formulas of ML_p. Then:

(i)   All atomic formulas belong to Per,

(ii) $A, B \in$ Per imply $\sim A$, $[A \to B] \in$ Per,

(iii) $A \in$ Per implies $\Box A \in$ Per,

(iv) $A \in$ Per implies $\forall x_e \, A \in$ Per,

(v)   Suppose $A \in$ Per and $F(x_\sigma)$ is an atomic formula of the form $s \, s^0 \ldots x \ldots s^{n-1}$ in which the variable x_σ occurs non-initially. Then the formulas $\forall x_\sigma \, [F(x) \to A]$ and $\exists x_\sigma \, [F(x) \land A]$ belong to Per.

From generalized completeness (Theorem 9.1) and the definition of persistence, we obtain

THEOREM 9.3. Let Γ and Σ be sets of persistent formulas, A a persistent formula of ML_p. Then:

(i) $\models A$ in ML_p if and only if $\vdash A$ in ML_p,

(ii) $\Gamma \models A$ in ML_p if and only if $\Gamma \vdash A$ in ML_p,

(iii) Σ is consistent in ML_p if and only if Σ is satisfiable in ML_p,

(iv) Σ is satisfiable in ML_p if and only if every finite subset Σ' of Σ is satisfiable in ML_p.

<u>Modal Predicate Logic with Comprehension</u>. Among the various axiomatic extensions of ML_p it is most natural to consider the deductive theory we denote by ML_p+C, obtained by adding to the axioms of ML_p all instances of the following <u>comprehension schema</u>:

$$C^{\sigma,A} : \quad \exists f_\sigma \Box \forall x^0 \forall x^1 \ldots \forall x^{n-1} [f_\sigma x^0 x^1 \ldots x^{n-1} \leftrightarrow A],$$

where $\sigma = (\sigma_0, \sigma_1, \ldots, \sigma_{n-1})$, x^k is of type σ_k for $k < n$, and f_σ is the first variable of type σ which is not free in the formula A.[4] This schema expresses the principle, valid in ML_p, that every formula with free variables determines a predicate, i.e., a relation-in-intension. A g-model of ML_p in which all instances $C^{\sigma,A}$ are true (i.e., satisfied by every index and assignment) is called a <u>general model (g-model) of</u> ML_p+C. It is evident that generalized completeness carries over to the logic ML_p+C.

<u>Extensional Comprehension</u>. It is reasonable to ask whether the ordinary comprehension principle, that every formula with free variables determines a relation, can also be expressed in the language of ML_p. Although the models of ML_p admit only predicates at the σth type level for each $\sigma \neq e$, we can identify ordinary relations with <u>constant</u> predicates, so that, e.g., a relation

$$R \subseteq M_{\sigma_0} \times \ldots \times M_{\sigma_{n-1}}$$

would be represented by the predicate $F \in M_\sigma$, $\sigma = (\sigma_0, \ldots, \sigma_{n-1})$, satisfying $F(i) = R$ for all $i \in I$. That the variable f_σ denotes such a constant predicate is expressible in ML_p by the formula

$$Rn(f) : \quad \forall x^0 \ldots \forall x^{n-1} [\Box f x^0 \ldots x^{n-1} \vee \Box \sim f x^0 \ldots x^{n-1}],$$

where x^k is of type σ_k for $k < n$. The principle of <u>extensional comprehension</u> is then expressed by the schema:

$$EC^{\sigma,A} : \quad \Box \exists f_\sigma [Rn(f) \wedge \forall x^0 \ldots \forall x^{n-1} [f x^0 \ldots x^{n-1} \leftrightarrow A]],$$

where $\sigma = (\sigma_0, \ldots, \sigma_{n-1})$, x^k is of type σ_k for $k < n$, and f_σ is the first variable of type σ which is not free in the formula A. We denote by ML_p+C+EC the theory obtained by adding all instances $EC^{\sigma,A}$ to the axioms of ML_p+C, and define a <u>general model (g-model) of</u> ML_p+C+EC in the obvious way. Note that a g-model M of ML_p is a g-model of ML_p+C

---

[4] The notation $C^{\sigma,A}$ was given a different meaning on page 71, when A is a formula of L_p. We shall refer to an <u>instance of the comprehension schema in L_p</u>, when it is necessary to distinguish the earlier formula from the present one.

just in case the following condition holds: For every $\sigma = (\sigma_0,\ldots,\sigma_{n-1})$, every formula $A(x^0,\ldots,x^{n-1})$ where x^k has type σ_k, and every assignment a over M, the predicate F belongs to M_σ, where

$$F(i) = \{ (X_0,\ldots,X_{n-1}) \mid M; i; a, X_0,\ldots,X_{n-1} \text{ sat } A \}$$

for each $i \in I$. If in addition the constant predicates G_i defined by $G_i(j) = F(i)$ ($j \in I$) always belong to M_σ, then M is a g-model of ML_p+C+EC. Hence:

LEMMA 9.4. The theory ML_p+C+EC is equivalent to the theory obtained by adding to the axioms of ML_p+C the formulas

E^σ : $\Box \forall f_\sigma \exists g_\sigma [Rn(g) \land f \equiv g]$

for every $\sigma \neq e$.

Some remarks about the schema EC are in order. It was discovered in the course of proving that the theories IL and ML_p have mutually interpretable extensions (Corollaries 13.6 and 13.12). Initially it seemed to the author that IL+D and ML_p+C would be equivalent theories in this sense, but it proved necessary to add the schema EC for the argument to go through. Although EC seems weaker than the more natural schema C of comprehension, we shall see in §15 that neither schema is stronger than the other, and in particular EC is independent of ML_p+C ; i.e., there exist g-models of ML_p+C in which EC fails. The discovery that EC is in fact a stronger principle than originally suspected apparently confirms a conjecture of Bressan,[5] who first made mention of an equivalent schema in his paper Bressan [1964]. We shall return to the schema EC in §11, where we introduce certain axioms of a rather different character which nevertheless prove to be equivalent to EC.

---

[5] Bressan [1972].

§10. Propositions in ML_p

Given an arbitrary g-model M of ML_p with index set I, we can define, for each formula A and assignment a, the _intension_ $Int_a[A]$ of A with respect to a; viz., we take it to be the unique proposition P in the set 2^I such that for $i \in I$, $P(i) = 1$ if M, i, a sat A, and $P(i) = 0$ otherwise. We have seen that the domain M_ϕ is always a non-empty set of propositions, which we call the _propositions of_ the g-model M. In general, the proposition $Int_a[A]$ determined by a formula and an assignment may fail to belong to M_ϕ; however, if M satisfies comprehension then in particular M, a satisfy

$$C^{\phi,A}: \quad \exists p_\phi \, \Box \, [\, p \leftrightarrow A \,] \; ,$$

where p_ϕ is the first variable of type ϕ which does not occur free in A, and from this it follows easily that $Int_a[A] \in M_\phi$.

The Algebra $B(M)$ _of Propositions_. Let M be a g-model of $ML_p + C$ with index set I, and let P be a proposition of M. We can identify P in the usual way with the subset X of I such that $i \in X$ if and only if $P(i) = 1$. Under this identification M_ϕ is put in one-to-one correspondence with a class of subsets of I, which we denote by $B(M)$.

THEOREM 10.1. Let M be a g-model of $ML_p + C$ with index set I. Then $B(M)$ is a subalgebra of the Boolean algebra of all subsets of I, which we call the _algebra of propositions of_ M.

Proof: $B(M)$ is non-empty, since M_ϕ is non-empty. If $P \in M_\phi$ then by comprehension (and rewrite of bound variables), $M; P$ satisfy the formula $\exists q_\phi \, \Box \, [q \leftrightarrow \sim p]$, so there exists $Q \in M_\phi$ with $Q(i) = 1$ if and only if $P(i) = 0$. Hence $B(M)$ is closed under complements. Similarly, if $P, Q \in M_\phi$ then $M; P, Q$ sat $\exists r_\phi \, \Box \, [r \leftrightarrow p \wedge q]$, so there exists $R \in M_\phi$ with $R(i) = 1$ if and only if $P(i) = Q(i) = 1$, and $B(M)$ is closed under intersections.

A subset X of I is called M-definable if there exist a formula A and an assignment a such that X consists of those $i \in I$ for which M, i, a sat A. Using Lemma 9.1.1 it is easily shown that:

THEOREM 10.2. Let M be a g-model of ML_p with index set I. Then the M-definable subsets of I form a Boolean algebra, and this algebra coincides with B(M) when M is a g-model of ML_p+C.

Indicial Equivalence. Let M be a g-model of ML_p, and let $i, j \in I$. We say that the index i is equivalent to the index j, and write $i \simeq j$, if for every formula A and assignment a, M, i, a sat A if and only if M, j, a sat A. Equivalently, $i \simeq j$ if and only if i and j belong to exactly the same M-definable subsets of I. The relation $\simeq$ is an equivalence relation on I, whose equivalence classes play a role analogous to that of the "sets of indiscernibles" of model theory.

THEOREM 10.3. Let M be a g-model of ML_p+C. Then for all $i, j \in I$ the following conditions are equivalent:

(i) $i \simeq j$,

(ii)   For every $\sigma \neq e$ and every $F \in M_\sigma$, $F(i) = F(j)$,

(iii)   For some $\sigma \neq e$ and every $F \in M_\sigma$, $F(i) = F(j)$,

(iv)   For every proposition P of M, $P(i) = P(j)$,

(v)   For every $X \in B(M)$, $i \in X$ if and only if $j \in X$.

Proof: By Theorem 10.2, (i) and (v) both assert that i and j belong to the same M-definable subsets of I, and are therefore equivalent. Clearly (iv) and (v) are equivalent, and (ii) implies (iv) implies (iii). We show that (iii) implies (iv) implies (ii). Assume $F(i) = F(j)$ for all $F \in M_\sigma$, where $\sigma = (\sigma_0, \ldots, \sigma_{n-1})$. Suppose $P \in M_\phi$; by comprehension,

$$M; P \text{ sat } \exists f_\sigma \Box \forall x^0 \ldots \forall x^{n-1} [f x^0 \ldots x^{n-1} \leftrightarrow p_\phi],$$

where p_ϕ is distinct from $f, x^0, \ldots, x^{n-1}$, so there exists $F \in M_\sigma$ such that, choosing $X_k \in M_{\sigma_k}$ arbitrarily for $k < n$, we have for every $i' \in I$: $(X_0, \ldots, X_{n-1}) \in F(i')$ if and only if $P(i') = 1$. Since $F(i) = F(j)$, this gives immediately $P(i) = P(j)$. Now assume (iv), and suppose $\sigma = (\sigma_0, \ldots, \sigma_{n-1})$, $F \in M_\sigma$, and $X_k \in M_{\sigma_k}$ for $k < n$. By comprehension,

$$M; F, X_0, \ldots, X_{n-1} \text{ sat } \exists p_\phi \square [p \leftrightarrow f_\sigma x^0 \ldots x^{n-1}],$$

where p_ϕ is not among f, x^0, ..., x^{n-1}, so that there exists $P \in M_\phi$ such that $P(i') = 1$ if and only if $(X_0, \ldots, X_{n-1}) \in F(i')$, for $i' \in I$. In particular, the sequence $(X_0, \ldots, X_{n-1})$ belongs to $F(i)$ just in case it belongs to $F(j)$, and since X_0, ..., X_{n-1} were arbitrary we conclude that $F(i) = F(j)$, proving (ii).

It should be observed that if M is a standard model then $M_\phi = 2^I$, so that $B(M)$ is the algebra $P(I)$ of all subsets of I. In this case the relation $\simeq$ is just the identity relation on I. In an arbitrary g-model of ML_p, or even ML_p+C, the relation $\simeq$ may not be the identity relation on I; a g-model M of ML_p is said to be <u>simple</u> if we have, for every i, $j \in I$: $i \simeq j$ if and only if $i = j$. Equivalently, M is simple if whenever $i \neq j$ in I there exist a formula A and assignment a such that M, i, a sat A but not M, j, a sat A. We now show that, in a precise sense, every g-model of ML_p can be replaced by a simple one.

<u>Indicial Homomorphisms</u>. Let M and M' be g-models of ML_p based on D, I and D', I' respectively. An <u>indicial homomorphism from</u> M <u>onto</u> M' is a family $\theta = (\vartheta, \vartheta_\sigma)_{\sigma \in P}$ of mappings such that:

(i) ϑ is a mapping from I onto I',

(ii)   For each $\sigma \in P$, ϑ_σ is a one-to-one mapping from M_σ onto M'_σ,

(iii)   For each $\sigma = (\sigma_0, \ldots, \sigma_{n-1})$, $F \in M_\sigma$, $i \in I$ and $X_k \in M_{\sigma_k}$ ($k < n$),
$(\vartheta_{\sigma_0}(X_0), \ldots, \vartheta_{\sigma_{n-1}}(X_{n-1})) \in \vartheta_\sigma(F)[\vartheta(i)]$ iff $(X_0, \ldots, X_{n-1}) \in F(i)$,

(iv)   For every constant c_σ, $m'(c_\sigma) = \vartheta_\sigma[m(c_\sigma)]$, where m and m' are the meaning functions of M and M' respectively.

If there exists an indicial homomorphism from M onto M' we say that M is <u>homomorphic</u> to M' and that M' is a <u>homomorphic image</u> of M. If the mapping ϑ is one-to-one, we say that θ is an <u>indicial isomorphism</u>, and that M and M' are <u>isomorphic</u>. Note that an indicial homomorphism θ is completely determined by ϑ and ϑ_e. The composition of two homomorphisms is again a homomorphism, and isomorphism is as usual an equivalence relation between g-models.

THEOREM 10.4. Let M and M' be g-models of ML_p, and let θ be an indicial homomorphism from M onto M'. For each $a \in As(M)$ let $\theta[a] \in As(M')$ be defined by $\theta[a](x_\sigma) = \vartheta_\sigma[a(x_\sigma)]$. Then for every formula A of ML_p, every $i \in I$ and $a \in As(M)$, we have

M, i, a sat A if and only if $M', \vartheta(i), \theta[a]$ sat A.

Proof: Clearly for every symbol s_σ we have $\overline{\theta[a]}(s_\sigma) = \vartheta_\sigma[\bar{a}(s_\sigma)]$. The proof proceeds by a routine induction on A.

COROLLARY 10.5. If θ is an indicial homomorphism from M onto M', and $\simeq$ and $\simeq'$ are the relations of indicial equivalence in M and M' respectively, then for all $i, j \in I$: $i \simeq j$ iff $\vartheta(i) \simeq' \vartheta(j)$.

Proof: By Theorem 10.4 and the definition of indicial equivalence.

Quotient G-Models. Let $M = (M_\sigma, m)_{\sigma \in P}$ be a g-model of ML_p based on D and I, and let ϑ be the canonical mapping from I onto the set $I/\simeq$ of equivalence classes of indices under the relation $\simeq$ in M. We define a quotient g-model

$$M/\simeq = (M_\sigma/\simeq, m/\simeq)_{\sigma \in P}$$

based on D and $I/\simeq$, and canonical one-to-one mappings ϑ_σ from M_σ onto $M_\sigma/\simeq$, as follows: We first put $M_e/\simeq = D = M_e$ and let ϑ_e be the identity mapping on D. For $\sigma = (\sigma_0, \ldots, \sigma_{n-1})$, we assume that $M_{\sigma_k}/\simeq$, ϑ_{σ_k} have already been defined for $k < n$, with ϑ_{σ_k} mapping M_{σ_k} one-to-one onto $M_{\sigma_k}/\simeq$. For each $F \in M_\sigma$ we define $\vartheta_\sigma(F)$ in the set

$$P(M_{\sigma_0}/\simeq \times \ldots \times M_{\sigma_{n-1}}/\simeq)^{I/\simeq}$$

by: $(\vartheta_{\sigma_0}(X_0), \ldots, \vartheta_{\sigma_{n-1}}(X_{n-1})) \in \vartheta_\sigma(F)[i/\simeq]$ iff $(X_0, \ldots, X_{n-1}) \in F(i)$, for any $X_0, \ldots, X_{n-1}$. This is well-defined, since in Theorem 10.3 it is easily checked that (i) implies (ii) in any g-model of ML_p. Clearly ϑ_σ is one-to-one on M_σ. We can therefore let $M_\sigma/\simeq$ be the range of ϑ_σ. Finally, for each constant c_σ we let $(m/\simeq)(c_\sigma) = \vartheta_\sigma[m(c_\sigma)]$.

THEOREM 10.6. Let M be a g-model of ML_p, $M/\simeq$ the quotient g-model defined above. Then $\theta = (\vartheta, \vartheta_\sigma)_{\sigma \in P}$ is an indicial homomorphism from M

PROPOSITIONS IN ML_p 83

onto $M/\simeq$. Moreover, $M/\simeq$ is simple.

Proof: By the construction and Corollary 10.5.

COROLLARY 10.7. Every g-model is homomorphic to a simple g-model.

Combining Theorems 10.4 and 10.6, we see that if M is a g-model of ML_p+C then $M/\simeq$ will also be a g-model of ML_p+C . Therefore:

COROLLARY 10.8. If Σ is a set of formulas of ML_p and Σ is g-satisfiable in ML_p (respectively, ML_p+C), then Σ is g-satisfiable in a simple g-model of ML_p (respectively, ML_p+C).

We also have:

COROLLARY 10.9. Let M be a g-model of ML_p. Then M is simple if and only if every indicial homomorphism on M is an isomorphism.

Proof: Theorem 10.6 and Corollary 10.5.

It should be remarked that the notion of a quotient g-model can be generalized. If M is a g-model of ML_p based on D and I , $\simeq$ is the relation of indicial equivalence in M , and $\cong$ is an equivalence relation on I for which $i \cong j$ implies $i \simeq j$, then the quotient g-model $M/\cong$ can be defined exactly as above. For this more general notion of quotient, analogues of the usual homomorphism theorems can be proved. Moreover, one can define similar notions of indicial equivalence, homomorphism and quotient for g-models of IL.

THEOREM 10.10. Let M' be a g-model of ML_p based on D' and I' , and let ϑ be an arbitrary mapping from a set I onto I' . Then there exists a g-model M of ML_p based on D' and I , and an indicial homomorphism θ from M onto M' extending ϑ .

Proof: Suppose $M' = (M'_\sigma, m')_{\sigma \in P}$. We define $M = (M_\sigma, m)_{\sigma \in P}$ and one-to-one mappings ϑ_σ from M_σ onto M'_σ , as follows: We first put $M_e = D'$ $= M'_e$ and let ϑ_e be the identity mapping on D' . For $\sigma = (\sigma_0, \ldots, \sigma_{n-1})$ we assume that M_{σ_k} and ϑ_{σ_k} are already defined for $k < n$, such that ϑ_{σ_k} maps M_{σ_k} one-to-one onto M'_{σ_k} . For each $F' \in M'_\sigma$ there exists a unique corresponding $F \in P(M_{\sigma_0} \times \ldots \times M_{\sigma_{n-1}})^I$ defined by the condition $(X_0, \ldots, X_{n-1}) \in F(i)$ if and only if $(\vartheta_{\sigma_0}(X_0), \ldots, \vartheta_{\sigma_{n-1}}(X_{n-1})) \in F'[\vartheta(i)]$. Moreover, the mapping of F' to F is clearly one-to-one, so we can let M_σ be its range and ϑ_σ its inverse. To complete the definition of M, we let $m(c_\sigma) \in M_\sigma$ be chosen so that $\vartheta_\sigma[m(c_\sigma)] = m'(c_\sigma) \in M'_\sigma$. It is easily verified that $\theta = (\vartheta, \vartheta_\sigma)_{\sigma \in P}$ is the desired homomorphism.

As remarked earlier, all standard models of ML_p are simple, although general models may not be. It follows from Theorems 10.4 and 10.10 that it is impossible to characterize the simple g-models of ML_p or ML_p+C by means of a new axiom or axioms. We are compensated, however, by the fact (Corollary 10.7) that we can always pass from a given g-model to its quotient, which is simple and satisfies exactly the same formulas.

We can characterize the simple g-models of ML_p+C in another way: Suppose M is a g-model of ML_p+C, and let $B(M)$ be the algebra of propositions of M. Then by Theorem 10.3 we see that M is simple if and only if $B(M)$ <u>separates points</u> in I; i.e., whenever $i \neq j$ in I, there exists a set $X \in B(M)$ with $i \in X$ and $j \notin X$.

§11. Atomic Propositions and EC

Suppose that M is a standard model of ML_p based on D and I. Then M is simple, and therefore $B(M)$ separates points in I; in fact, if $i \neq j$ in I then $\{i\}$ separates i and j and belongs to $B(M)$, since $B(M)$ contains every subset of I. This has the interesting consequence that, in a standard model, there exists for each index i a strongest proposition which is true at i, viz., the proposition P which is true at i and false at every $j \neq i$. If Q is any other proposition true at i, then P <u>strictly implies</u> Q, in the sense that Q is true at j whenever P is true at j. Consequently, the formula

At: $\qquad \Box \exists p_\phi [p \wedge \forall q_\phi [q \rightarrow \Box [p \rightarrow q]]]$

which expresses the principle that there necessarily exists a strongest true proposition, is valid in ML_p, i.e., true in all standard models.

There are closely related conditions which we might also consider. Let us call a proposition P <u>atomic</u> if (i) P is possibly true, and (ii) for every proposition Q, P strictly implies either Q or its negation. This can be expressed by the formula:

$$\Diamond p_\phi \land \forall q_\phi \; [\, \Box \, [p \to q] \lor \Box \, [p \to \sim q] \,] \;,$$

which we abbreviate by $\text{Atom}(p_\phi)$. The formulas

At_1 : $\Box \, \exists p_\phi \, [\, \text{Atom}(p) \land p \,] \;,$

At_2 : $\forall p_\phi \, [\, \Diamond p \to \exists q_\phi [\, \text{Atom}(q) \land \Box \, [q \to p] \,] \,] \;,$

then express the respective principles that (1) there necessarily exists a true atomic proposition, and (2) every possibly true proposition is strictly implied by an atomic proposition. Both At_1 and At_2 are valid in ML_p, and clearly we have:

LEMMA 11.1. Let M be a g-model of ML_p+C with index set I, and let B(M) be the algebra of propositions of M. Then:

(i)   M sat At if and only if for every $i \in I$ there is a smallest set $X \in B(M)$ for which $i \in X$,

(ii)   M sat At_1 if and only if every $i \in I$ belongs to an atom in the Boolean algebra B(M),

(iii) M sat At_2 if and only if B(M) is atomic.

THEOREM 11.2. The formulas At, At_1, At_2 are provably equivalent in ML_p+C.

Proof: It is easily verified that for any algebra B of subsets of a set I, the conditions (i) For every $i \in I$ there is a smallest set X in B for which $i \in X$, and (ii) Every $i \in I$ belongs to an atom of B, are equivalent, and both imply the condition (iii) B is atomic. By Lemma 11.1 and generalized completeness, the formulas [At $\leftrightarrow$ At_1] and [$\text{At}_1 \to \text{At}_2$] are therefore provable in ML_p+C. Although (iii) does not imply (i) for an arbitrary field B of sets, we can still prove the implication [$\text{At}_2 \to$ At] in the theory ML_p+C.[1] For, suppose M is a g-model

---

[1] Cf. Fine [1970], p. 341.

of ML_p+C with index set I, and M sat At_2. Then $B(M)$ is atomic, by Lemma 11.1, and it suffices to show that every index $i \in I$ belongs to a smallest set $X \in B(M)$, or equivalently that every $i \in I$ belongs to an atom in $B(M)$. By comprehension,

$$M \text{ sat } \exists p_\phi \Box [p \leftrightarrow \sim \exists q_\phi [Atom(q) \wedge q]],$$

from which it follows that there exists a set $X_0 \in B(M)$ such that $i \in X_0$ just in case i belongs to no atom of $B(M)$. Thus, if some i belongs to no atom then $X_0 \neq \phi$, and therefore X_0 dominates some atom Y. Since $Y \neq \phi$ we can choose $i \in Y$; but then $i \in X_0$, contradicting the definition of X_0.

We refer to the formula At as the <u>axiom of atomic propositions</u>, and we denote by ML_p+C+At the theory obtained by adding At to the axioms of ML_p+C. A <u>general model (g-model) of</u> ML_p+C+At is defined accordingly. Axiom At originates with Kaplan [1970], who considers an extension S5Q of the usual propositional modal logic S5 in which quantifiers over propositional variables are permitted, and gives an axiomatization which is complete for the (standard) possible world semantics. The formula At appears as Axiom 8 in his formulation, and he remarks that it is independent of the other axioms. In §15 we prove that At is also independent of ML_p+C, a considerably stronger theory than S5Q.[2] Axiom At also appears in the logic S5π+ of Fine [1970], which is almost identical with Kaplan's S5Q.

Before proving the main result of the present section, we have the following

LEMMA 11.3. Let M be a g-model of ML_p+C with index set I, and let $\simeq$ be the relation of indicial equivalence in M. Then for each index $i \in I$, the following conditions are equivalent:

(i) The equivalence class $i/\simeq$ belongs to $B(M)$,

(ii) $i/\simeq$ is the unique atom of $B(M)$ containing i,

(iii) i belongs to an atom of $B(M)$.

---

[2] Kaplan's independence proof, which is based on a normal form theorem for S5Q, does not seem to generalize to ML_p+C. The Boolean methods employed in §15, however, apply equally well to S5Q.

ATOMIC PROPOSITIONS AND EC   87

Proof: Assume (i). Then by Theorem 10.3 we have either $[i/\simeq] \subseteq X$ or else $[i/\simeq] \cap X = \phi$ for every $X \in B(M)$, from which it follows that $i/\simeq$ is an atom of $B(M)$ containing i, and clearly such an atom must be unique. Therefore (ii) holds. Trivially (ii) implies (iii). Assume (iii); say i belongs to the atom X_0 of $B(M)$. By Theorem 10.3, $[i/\simeq] \subseteq X_0$, so it suffices to show $X_0 \subseteq [i/\simeq]$. Suppose $j \in X_0$; then clearly i, j belong to exactly the same elements X of $B(M)$, whence by Theorem 10.3 again, $i \simeq j$ and therefore $j \in [i/\simeq]$, as desired.

COROLLARY 11.4. Let M be a g-model of ML_p+C with index set I. Then M sat At if and only if $i/\simeq$ belongs to $B(M)$ for all $i \in I$. Hence, if M is simple then M sat At if and only if $B(M)$ contains all singletons $\{i\}$ for $i \in I$.

We can now prove:

THEOREM 11.5. The theories ML_p+C+EC and ML_p+C+At are equivalent.

Proof: In view of Lemma 9.4 it is sufficient to show that the theory ML_p+C+At is equivalent to the theory obtained by adding to the axioms of ML_p+C all the formulas E^σ for $\sigma \neq e$. The next two lemmas actually show somewhat more. For each $n \in \omega$, let $\bar{n}$ denote the n-tuple $(e,e,\ldots,e) \in P$, so that in particular $\bar{0}$ is the type ϕ.

LEMMA 11.5.1. $[At \to E^\sigma]$ is provable in ML_p+C for $\sigma \neq e$.

LEMMA 11.5.2. $[E^\sigma \to At]$ is provable in ML_p+C for $\sigma \neq e, \bar{n}$ ($n \in \omega$).

Proof of 11.5.1: We use generalized completeness. Let $M = (M_\sigma, m)_{\sigma \in P}$ be a g-model of ML_p+C with index set I, and assume that M sat At. We show that M satisfies

E^σ : $\square \forall f_\sigma \exists g_\sigma [Rn(g) \wedge f \equiv g]$,

where $\sigma = (\sigma_0,\ldots,\sigma_{n-1})$ and $[f \equiv g]$ abbreviates the formula

$$\forall x^0 \ldots \forall x^{n-1} [f x^0 \ldots x^{n-1} \leftrightarrow g x^0 \ldots x^{n-1}].$$

Suppose $i \in I$, $F \in M_\sigma$; we shall find $G \in M_\sigma$ for which $M; i; F,G$ sat $[Rn(g) \wedge f \equiv g]$. Since M sat At, there exists $P \in M_\phi$ such that $\{ j \mid P(j) = 1 \}$ is an atom of $B(M)$ containing i. By Lemma 11.3, we have $P(j) = 1$ if and only if $i \simeq j$, for all $j \in I$. Now by comprehension in M (rewriting bound variables), $M; P, F$ satisfy the formula

$$\exists g_\sigma \square \forall x^0 \ldots \forall x^{n-1} [g x^0 \ldots x^{n-1} \leftrightarrow \Diamond [p_\phi \wedge f_\sigma x^0 \ldots x^{n-1}]],$$

where x^k is of type σ_k. Hence there exists $G \in M_\sigma$ such that for all $i' \in I$ and $X_k \in M_{\sigma_k}$ ($k < n$) we have $(X_0, \ldots, X_{n-1}) \in G(i')$ if and only if $(X_0, \ldots, X_{n-1}) \in F(j)$ for some j such that $P(j) = 1$. But $P(j) = 1$ just in case $i \simeq j$, so by Theorem 10.3 it follows that $(X_0, \ldots, X_{n-1})$ belongs to $G(i')$ if and only if it belongs to $F(i)$. From this we immediately have $M; i; F, G$ sat $[Rn(g) \wedge f \equiv g]$, and the proof is complete.

We remark that E^ϕ is itself provable in ML_p+C, as is easily seen. However, the formula E^σ is not provable in ML_p+C for $\sigma \neq e$, ϕ, as we show in §15.

Proof of 11.5.2: Let σ be a type different from e and $\bar{n}$ for all $n \in \omega$. Then $\sigma = (\sigma_0, \ldots, \sigma_k, \ldots, \sigma_n)$ where $\sigma_k = (\tau_0, \ldots, \tau_{m-1})$. We use generalized completeness to show that the formula $[E^\sigma \to At]$ is provable in ML_p+C. Let $M = (M_\sigma, m)_{\sigma \in \rho}$ be a g-model of ML_p+C with index set I which satisfies E^σ. We show that M sat At. Suppose $i \in I$. By comprehension, M satisfies the formula

$$\exists f_\sigma \square \forall x^0 \ldots \forall x^n [f x^0 \ldots x^n \leftrightarrow \exists y^0 \ldots \exists y^{m-1} x^k y^0 \ldots y^{m-1}]],$$

where x^ℓ is of type σ_ℓ for $\ell \leq n$, and $y^0, \ldots, y^{m-1}$ are the first distinct variables of types $\tau_0, \ldots, \tau_{m-1}$, respectively. Therefore there exists $F \in M_\sigma$ such that for all $j \in I$ and $X_\ell \in M_{\sigma_\ell}$ ($\ell \leq n$) we have $(X_0, \ldots, X_n) \in F(j)$ if and only if $X_k(j) \neq \phi$. Since M satisfies E^σ, we obtain $G \in M_\sigma$ for which $G(j) = F(i)$ for every $j \in I$. Thus, for all $j \in I$ and $X_\ell \in M_{\sigma_\ell}$ ($\ell \leq n$) we have $(X_0, \ldots, X_n) \in G(j)$ if and only if $X_k(i) \neq \phi$. Now by comprehension, $M; G$ satisfy the formula

$$\exists p_\phi \square [p \leftrightarrow \forall x^0 \ldots \forall x^n [g_\sigma x^0 \ldots x^n \to \exists y^0 \ldots \exists y^{m-1} x^k y^0 \ldots y^{m-1}]],$$

from which it follows that there exists $P \in M_\phi$ such that for all $j \in I$, $P(j) = 1$ if and only if for every $X \in M_{\sigma_k}$, $X(i) \neq \phi$ implies $X(j) \neq \phi$. Clearly $P(i) = 1$, so it remains only to show that $M; i; P$ satisfy

$\forall q_\phi \ [q \to \Box \ [p_\phi \to q]]$. Suppose $Q \in M_\phi$, $Q(i) = 1$. We must show that $Q(j) = 1$ whenever $P(j) = 1$. By comprehension,

$$M; Q \text{ sat } \exists x_{\sigma_k} \ \Box \ \forall y^0 \ldots \forall y^{m-1} \ [\ x \ y^0 \ldots y^{m-1} \ \leftrightarrow \ q_\phi \ ] \ ,$$

where q_ϕ is the first variable of type ϕ distinct from x , y^0 , ... , y^{m-1} . Thus, there exists $X \in M_{\sigma_k}$ such that for all $j \in I$ and $Y_\ell \in M_{\tau_\ell}$ ($\ell < m$) we have $(Y_0, \ldots, Y_{m-1}) \in X(j)$ if and only if $Q(j) = 1$, and therefore for all $j \in I$, $X(j) \neq \phi$ if and only if $Q(j) = 1$. But we have $Q(i) = 1$, so $X(i) \neq \phi$, and hence for all $j \in I$, $P(j) = 1$ implies $X(j) \neq \phi$, which implies $Q(j) = 1$. Thus we have $Q(j) = 1$ whenever $P(j) = 1$, and the proof is complete.

Theorem 11.5 shows that instead of adding to ML_p+C the axiom schema EC of extensional comprehension, we can equivalently add the single axiom At of atomic propositions. We return to consider various independence questions related to these theories in Chapter 4.

§12. Propositional Operators

Montague [1970a] outlines a general treatment of one-place propositional operators within his formalized <u>Pragmatics</u>, and shows how such operators can be interpreted as properties of propositions. In this section we develop this idea, using the fact that we can express in ML_p various properties of these operators. In particular, we shall see that we can accommodate within ML_p+C modal operators satisfying various of the Lewis axiom systems, even though ML_p+C itself is based on an S5 modality.

M-<u>Formulas</u>. For the purposes of this section (and again in Chapter 4) we find it notationally convenient to extend the semantics of ML_p in the following way: Let $M = (M_\sigma, m)_{\sigma \in P}$ be a g-model of ML_p based on D and I ; we wish to add to the vocabulary of ML_p new constant symbols to act as names of the various elements $X \in M_\sigma$ for $\sigma \in P$. For simplicity, let us take the object X as a name for itself; i.e., we add X itself as a new constant of type σ whenever $X \in M_\sigma$, and we extend the meaning function m of M by letting $m(X) = X$.[1] A formula of this extended language (which will in general have a non-denumerable vocabulary) will be called an M-<u>formula</u>, and an M-<u>sentence</u> if it has no free variables. For an M-formula A , an index i , and an assignment a over M , the notion

$$M, i, a \text{ sat } A$$

is defined exactly as in §9, but taking into account the new constants. If $A(x_\sigma)$ is an M-formula containing the variable x_σ free, and c_σ is any constant of the extended language, it is easily shown[2] that

(*)   M, i, a sat $A(c)$ if and only if M; i; a,X sat $A(x)$,

where $X = m(c)$. It follows that the notion M, i sat A , where A is an M-sentence, can be defined directly by recursion on the length of A ; at the quantifier clause we simply stipulate that M, i sat $\forall x_\sigma A(x)$ if and only if M, i sat $A(X)$ for every $X \in M_\sigma$. We can therefore eliminate any reference to assignments by working with M-sentences instead of formulas of ML_p. Note that every M-formula has the form $A(X_0, \ldots, X_{n-1})$, where $A(x^0, \ldots, x^{n-1})$ is an ordinary formula of ML_p, $x^0, \ldots, x^{n-1}$ are distinct variables of types $\sigma_0, \ldots, \sigma_{n-1}$ respectively, and $X_k \in M_{\sigma_k}$ for $k < n$. By (*), therefore, we may think of

$$M, i, a \text{ sat } A(X_0, \ldots, X_{n-1})$$

as abbreviating the equivalent condition

$$M; i; a, X_0, \ldots, X_{n-1} \text{ sat } A(x^0, \ldots, x^{n-1}) .$$

<u>Propositional Operators of</u> M . Let $M = (M_\sigma, m)_{\sigma \in P}$ be a fixed g-model of $ML_p + C$, with index set I . An element F of $M_{(\phi)}$ is called a <u>propositional operator of</u> M . Since $M_{(\phi)} \subseteq P(M_\phi)^I$, we see that such op-

---

[1] Strictly speaking we should choose, for each $\sigma \in P$ and $X \in M_\sigma$, some new object c_σ^X which is not already a symbol of ML_p, in such a way that the mapping of (σ, X) to c_σ^X is one-to-one. We ignore these difficulties.

[2] Cf. Lemma 9.1.1.

erators are always properties of propositions of M.[3] Every M-formula $A(p_\phi)$, with at most the variable p_ϕ free, determines a unique operator of M; for by comprehension, M satisfies the M-sentence

$$\exists f_{(\phi)} \, \Box \, \forall p_\phi \, [\, fp \leftrightarrow A(p) \,] \, ,$$

and consequently M sat $\Box \, \forall p_\phi \, [Fp \leftrightarrow A(p)]$ for some $F \in M_{(\phi)}$; i.e., we have $P \in F(i)$ if and only if M, i sat $A(P)$, for every $P \in M_\phi$, $i \in I$. In particular, we always have the necessity and possibility operators of M, defined by:

$$\Box \, \forall p_\phi \, [\, F_\Box \, p \leftrightarrow \Box \, p \,] \, ,$$

$$\Box \, \forall p_\phi \, [\, F_\Diamond \, p \leftrightarrow \Diamond \, p \,] \, .$$

If $s_{(\phi)}$ is any symbol of type (ϕ) and A is any M-formula which is not a symbol of type ϕ standing alone, we introduce the abbreviation

$$sA \quad \text{for} \quad \exists p_\phi \, [\, \Box \, [p \leftrightarrow A] \wedge sp \,] \, ,$$

where p_ϕ is the first variable of type ϕ which does not occur free in A. Using generalized completeness it is easily shown that:

LEMMA 12.1. For any formulas A, B of ML_p and any variable $f_{(\phi)}$ of type (ϕ), the formula

$$\Box \, [A \leftrightarrow B] \rightarrow [\, fA \leftrightarrow fB \,]$$

is provable in $ML_p + C$.

In a g-model M of $ML_p + C$, therefore, it follows that for any index i, M, i sat $[\, \Box \, [A \leftrightarrow B] \rightarrow [FA \leftrightarrow FB] \,]$, whenever A and B are M-sentences and F is a propositional operator of M. In fact, by comprehension we can define, as in §10, the <u>intension</u> $Int[A]$ of an M-sentence A as the unique $P \in M_\phi$ for which M sat $\Box \, [P \leftrightarrow A]$; i.e., for which we have, for all $i \in I$: $P(i) = 1$ if and only if M, i sat A. It then follows that:

---

[3] Here we identify M_ϕ with the Cartesian product $M_{\sigma_0} \times \ldots \times M_{\sigma_{n-1}}$, where $n = 1$ and $\sigma_0 = \phi$, although these sets are slightly different.

LEMMA 12.2. Let M be a g-model of ML_p+C, A an M-sentence, F a propositional operator of M, $i \in I$. Then M, i sat FA if and only if $Int[A] \in F(i)$.

In a particular g-model M there may be various interesting propositional operators in addition to the modal operators defined earlier. <u>Tenses</u> provide a natural example: Let M be a standard model of ML_p whose index set I is the set of real numbers, thought of as moments in time. For an M-sentence A let M, i sat A express the intuitive condition that A is true in M at time i. Then we can define the <u>past tense</u> operator $F \in M_{(\phi)} = P(M_\phi)^I$ by letting $P \in F(i)$ just in case $P(j) = 1$ for some $j < i$. From Lemma 12.2 we see that for any M-sentence A, any $i \in I$, we have M, i sat FA if and only if M, j sat A for some $j < i$. Thus, FA may be given the reading "It has been the case that A." Other tenses can be treated similarly as propositional operators.

<u>Other Modalities</u>. We shall be interested in various systems of propositional modal logic, well-known from the literature.[4] Consider a language appropriate to propositional modal logic, in which formulas are built up from propositional variables $p, q, r \ldots$ by means of the sentential connectives $\sim, \to$ and the formal propositional operator N. Each of the modal calculi we consider takes its axiom schemata from among the following:

AS1. A, if A is tautologous in $\sim, \to$,

AS2. $N[A \to B] \to [NA \to NB]$,

AS3. $NA \to A$,

AS4. $A \to N \sim N \sim A$,

AS5. $NA \to NNA$,

AS6. $\sim NA \to N \sim NA$,

and has as its inference rules:

R1.   From A and $[A \to B]$ to infer B,

R2.   From A to infer NA.

---

[4] See Hughes and Cresswell [1968].

The systems we consider are Kripke's system[5] K, the Gödel-Feys-von Wright system T, the Brouwersche system B, and the Lewis systems S4 and S5. K contains the axiom schemata AS1 and AS2 alone, and is contained in the other systems. In addition, T contains AS3, B contains AS3 and AS4, S4 contains AS3 and AS5, and S5 contains AS3 and AS6 (or equivalently, AS3, AS4 and AS5). For each of these systems a natural semantics has been provided by Kripke, based on so-called "relevance relations" between indices.[6] Specifically, we take a <u>model of</u> K to be a pair $M = (I, R)$, where I is a non-empty set and R is a binary relation on I, and define an <u>assignment over</u> M to be a function a on the set of variables such that $a(p) \in 2^I$ for each variable p. We then define M, i, a sat A in the usual way, with the following clause for the modal operator: M, i, a sat NA if and only if M, j, a sat A whenever $i R j$. A formula A is <u>true in</u> M if M, i, a sat A for every i and a. A model $M = (I, R)$ of K is called a <u>model of</u> T if the relation R is reflexive on I, a <u>model of</u> B (resp., S4) if in addition R is symmetric (resp., transitive), and a <u>model of</u> S5 if R is an equivalence relation on I. Kripke [1963a] proved that a formula A is a theorem of K (resp., T, B, S4, S5) just in case A is true in every model of K (resp., T, B, S4, S5).

Corresponding to the axiom schemata AS2 through AS6 and the inference rule R2, we introduce the following formulas of ML_p, in which the variable $f_{(\phi)}$ occurs free:

$A_2(f)$: $\quad \Box \forall p_\phi \forall q_\phi [f[p \to q] \to [fp \to fq]]$,

$A_3(f)$: $\quad \Box \forall p_\phi [fp \to p]$,

$A_4(f)$: $\quad \Box \forall p_\phi [p \to f \sim f \sim p]$,

$A_5(f)$: $\quad \Box \forall p_\phi [fp \to ffp]$,

$A_6(f)$: $\quad \Box \forall p_\phi [\sim fp \to f \sim fp]$,

$R_2(f)$: $\quad \forall p_\phi [\Box p \to \Box fp]$.

---

[5] So designated in Kaplan [1966], p. 121. See Kripke [1963a], p. 95.

[6] Kripke [1963a]. The idea of using relevance relations was suggested earlier by Montague [1960], Kanger [1957], and Hintikka [1961]. These authors had in mind relations between models, however, in contrast to the indicial approach of Kripke.

Suppose that $M = (M_\sigma, m)_{\sigma \in P}$ is a g-model of ML_p+C with index set I, and let N be a propositional operator of M. N is called a K-<u>operator</u> if M satisfies the M-sentences $A_2(N)$ and $R_2(N)$; a T-<u>operator</u> (resp., B-<u>operator</u>, S4-<u>operator</u>, S5-<u>operator</u>) if in addition M satisfies $A_3(N)$ (resp., $A_3(N)$ and $A_4(N)$, $A_3(N)$ and $A_5(N)$, $A_3(N)$ and $A_6(N)$). To see the relationship between these operators and the corresponding modal calculi, suppose that, e.g., N is a K-operator of a g-model M of ML_p+C. Then for any M-formulas A and B, the M-formulas

(1) A, if A is tautologous in $\sim$, $\to$,

(2) $N[A \to B] \to [NA \to NB]$

will be true in M (i.e., satisfied by every i and a), and in addition

(1')   If A and $[A \to B]$ are true in M then B is true in M,

(2')   If A is true in M then NA is true in M.

Thus, any M-formula which is an instance (in the language of the g-model M) of a theorem of K will be true in M. Similar remarks apply to T-operators, B-operators, etc.

The propositional operators arising from relevance relations on the set I are of course of a special type. We can formally characterize such operators in ML_p; specifically, an operator N of a g-model M of ML_p+C is <u>indicial</u> if

$$M \text{ sat } \Box \exists p_\phi \forall q_\phi [Nq \leftrightarrow \Box [p \to q]].$$

Suppose this condition holds. Then N determines a binary relation R_N on the index set I of M, as follows: For each $i \in I$, let P_i be the unique $P \in M_\phi$ for which M, i sat $\forall q_\phi [Nq \leftrightarrow \Box [P \to q]]$. (That P_i is unique follows from the observation that M, i sat NP_i.) We define R_N by letting $i R_N j$ if and only if $P_i(j) = 1$. This relation is called the <u>relevance relation for</u> N, in view of the following straightforward

LEMMA 12.3. Let M be a g-model of ML_p+C with index set I, and let N be an indicial operator of M. Then for every M-sentence A and every $i \in I$: M, i sat NA if and only if M, j sat A whenever $i R_N j$.

COROLLARY 12.4. Let M be a g-model of ML_p+C. Then every indicial operator of M is a K-operator.

For indicial operators we can show that the axioms of the various modal calculi characterize exactly the corresponding properties of the relevance relation. Precisely:

THEOREM 12.5. Let M be a g-model of ML_p+C with index set I, and let N be an indicial operator of M. Then:

(i) N is a T-operator iff R_N is reflexive on I,

(ii) N is a B-operator iff R_N is reflexive and symmetric,

(iii) N is an S4-operator iff R_N is reflexive and transitive,

(iv) N is an S5-operator iff R_N is an equivalence relation on I.

Proof: We prove (ii); the proofs of (i), (iii) and (iv) are similar. First, if R_N is reflexive and symmetric we must verify that N is a B-operator. But this just follows Kripke's argument that every theorem of B is true in every model of B, in view of Lemma 12.3. For the converse, we assume that N is an indicial B-operator, so that M satisfies $A_2(N)$, $R_2(N)$, $A_3(N)$ and $A_4(N)$. To see that R_N is reflexive, let $i \in I$. Since N is indicial, we have $P_i \in M_\phi$, and clearly M, i sat NP_i, so using $A_3(N)$ we obtain M, i sat P_i, which implies $P_i(i) = 1$, i.e., $i\, R_N\, i$. To see that R_N is symmetric, suppose that $i\, R_N\, j$ but not $j\, R_N\, i$. Then $P_j(i) = 0$, i.e., M, i sat $\sim P_j$. By comprehension there exists $Q \in M_\phi$ such that M sat $\Box [Q \leftrightarrow \sim P_j]$, so that M, i sat Q. Using $A_4(N)$, it follows that M, i sat $N \sim N \sim Q$. But $i\, R_N\, j$, so by Lemma 12.3 M, j sat $\sim N \sim Q$, i.e., it is not the case that M, j sat $N \sim Q$. But this contradicts Lemma 12.3, since for all $i' \in I$, $j\, R_N\, i'$ implies $P_j(i') = 1$, which implies $Q(i') = 0$, i.e., M, i' sat $\sim Q$.

It is natural to ask whether the converse to Corollary 12.4 holds; i.e., whether every K-operator is indicial. It is easy to see, however, that this is not the case. In fact, we can give an example of an S4-operator in a standard model of ML_p which is not indicial. The example is the *present progressive* tense of Scott: Let I be the set of real numbers, and let M be a standard model of ML_p with index set I. Define the propositional operator $N \in M_{(\phi)} = P(M_\phi)^I$ by putting $P \in N(i)$ just in case

$P(j) = 1$ for all j in some open interval around i. If we think of the indices as moments in time, then for any M-sentence A, NA can be given the reading "It is being the case that A." It is easily checked using Lemma 12.2 that N is an S4-operator, but clearly N is not indicial. We shall see in §15 that some g-models of ML_p+C even contain non-indicial S5-operators. However:

THEOREM 12.6. In any g-model of ML_p+C+EC, every S5-operator is indicial.

COROLLARY 12.7. In any standard model of ML_p, every S5-operator is indicial.

Proof of 12.6: Let M be a g-model of ML_p+C+EC with index set I, and let N be an S5-operator of M, so that M satisfies $A_2(N)$, $R_2(N)$, $A_3(N)$ and $A_6(N)$. From the usual proof that S5 extends S4 and B, we conclude that M also satisfies $A_4(N)$ and $A_5(N)$. We show that

$$M \text{ sat } \Box \exists p_\phi \forall q_\phi [Nq \leftrightarrow \Box [p \rightarrow q]].$$

Suppose $i \in I$; we must find $P \in M_\phi$ for which

(*) M, i sat $\forall q_\phi [Nq \leftrightarrow \Box [P \rightarrow q]]$.

By Theorem 11.5, M sat At, so there exists $P' \in M_\phi$ for which M, i sat $P' \land \forall q_\phi [q \rightarrow \Box [P' \rightarrow q]]$, or equivalently, M, i sat $\forall q_\phi [q \leftrightarrow \Box [P' \rightarrow q]]$. By comprehension we therefore have

(1) M, i sat $[A \leftrightarrow \Box [P' \rightarrow A]]$

for every M-sentence A. Also by comprehension, there exists $P \in M_\phi$ such that M sat $\Box [P \leftrightarrow \forall q_\phi [\Box [P' \rightarrow Nq] \rightarrow q]]$. This together with (1) implies that for every $j \in I$:

(2) $P(j) = 1$ iff for every $Q \in M_\phi$, $Q \in N(i)$ implies $Q(j) = 1$.

We show that (*) holds for P. Suppose $Q \in M_\phi$, and assume first that M, i sat NQ, i.e., $Q \in N(i)$. Then from (2), we have $Q(j) = 1$ whenever $P(j) = 1$, so M sat $\Box [P \rightarrow Q]$. On the other hand, assume that M sat $\Box [P \rightarrow Q]$; then by $A_2(N)$, $R_2(N)$ and comprehension we easily obtain first that M sat $\Box N[P \rightarrow Q]$ and then that M sat $\Box [NP \rightarrow NQ]$.

Thus, if we can show

(**)   M, i sat NP ,

then we can conclude that M, i sat NQ , and the proof will be complete. We first show

(†)   M sat $\Box [\sim P \to N \sim P]$.

For, suppose M, j sat $\sim P$. Then by (2), there exists $Q_0 \in N(i)$ with $Q_0(j) = 0$, i.e., M, j sat $\sim Q_0$. Hence using $A_3(N)$, M, j sat $\sim NQ_0$. But M, i sat NQ_0 , so using $A_5(N)$ we conclude that M, i sat NNQ_0 . By comprehension, let $Q_1 \in M_\phi$ be such that M sat $\Box [Q_1 \leftrightarrow NQ_0]$. Then $Q_1 \in N(i)$, whence by (2), $P(i') = 1$ implies $Q_1(i') = 1$ for all i' $\in$ I, i.e., M sat $\Box [P \to NQ_0]$. It follows that M sat $\Box [\sim NQ_0 \to \sim P]$, so using $A_2(N)$, $R_2(N)$ and comprehension we obtain M sat $\Box [N \sim NQ_0 \to N \sim P]$. But using $A_6(N)$ we see that M sat $\Box [\sim NQ_0 \to N \sim NQ_0]$. Therefore M sat $\Box [\sim NQ_0 \to N \sim P]$, and since M, j sat $\sim NQ_0$ we conclude that M, j sat $N \sim P$. This completes the proof of (†).

In view of $A_3(N)$, (†) implies that M sat $\Box [\sim P \leftrightarrow N \sim P]$, or equivalently,

(3)   M sat $\Box [P \leftrightarrow \sim N \sim P]$,

which by Lemma 12.1 yields

(4)   M sat $\Box [NP \leftrightarrow N \sim N \sim P]$.

But by (2), using $A_3(N)$, we clearly have $P(i) = 1$, whence by (3), we have M, i sat $\sim N \sim P$, and therefore using $A_6(N)$ and comprehension, M, i sat $N \sim N \sim P$. This with (4) yields M, i sat NP , so that (**) holds and the theorem is proved.

We have seen that various classes of modal operators -- e.g., those obeying specified modal axioms, or those arising from relevance relations between indices -- can be characterized in a natural way by means of formal conditions expressible in ML_p. It would be interesting to know to what further extent the language of ML_p can be used in classifying propositional operators.

§13. Relative Strength of IL and ML_p

We now compare the logics IL and ML_p by means of respective translations of the formulas of each language into formulas of the other. In each case we have the expected result that the translation preserves the standard semantics: A formula of IL is valid in IL if and only if its translate is valid in ML_p, and vice-versa. However, these translations do not preserve the deductive theories IL and ML_p, or equivalently, the generalized semantics for these logics; in particular, there are theorems of IL whose translates are not theorems of ML_p. We therefore consider as well the extended theories IL+D and ML_p+C+EC , for which we prove strong relative interpretability, in the following sense: A formula is provable in one of these theories if and only if its translate is provable in the other.

<u>Interpretability of</u> ML_p <u>in</u> IL. For each $\sigma \in P$ we define a corresponding type $\alpha[\sigma] \in T$ as follows:

(i) $\alpha[e] = e$,

(ii) $\alpha[\sigma] = (s,(\alpha[\sigma_0],(\ldots(\alpha[\sigma_{n-1}],t)\ldots)))$ when $\sigma = (\sigma_0,\ldots,\sigma_{n-1})$.

To each symbol s_σ of ML_p we make correspond a symbol $\underline{s}_{\alpha[\sigma]}$ of IL:

(i)   If s is x^n_σ then $\underline{s}$ is $x^n_{\alpha[\sigma]}$,

(ii) If s is c^n_σ then $\underline{s}$ is $c^n_{\alpha[\sigma]}$.

For each formula A of ML_p we define a translate $\underline{A}$ in IL:

(i)   If A is $s_\sigma s^0 \ldots s^{n-1}$ then $\underline{A}$ is $\check{\underline{s}}\, \underline{s}^0 \ldots \underline{s}^{n-1}$,

(ii)   If A is $[s_e \equiv s'_e]$ then $\underline{A}$ is $[\,\underline{s} \equiv \underline{s}'\,]$,

(iii) If A is $\sim B$ then $\underline{A}$ is $\sim \underline{B}$,

(iv)   If A is $[B \to C]$ then $\underline{A}$ is $[\,\underline{B} \to \underline{C}\,]$,

(v)   If A is $\forall x_\sigma B$ then $\underline{A}$ is $\forall \underline{x}\, \underline{B}$,

(vi)   If A is $\Box B$ then $\underline{A}$ is $\Box \underline{B}$.

If Σ is a set of formulas of ML_p we denote by $\underline{\Sigma}$ the set of all translates $\underline{A}$ for $A \in \Sigma$. Also, we let Δ_Σ consist of all formulas of IL of the form $\exists x_{\alpha[\sigma]} [\underline{c} \equiv x]$, where c_σ is a constant occurring in Σ.

THEOREM 13.1. Let Σ be a set of formulas of ML_p. Then Σ is satisfiable in ML_p if and only if $\underline{\Sigma} \cup \Delta_\Sigma$ is satisfiable in IL.

Proof: We prove one implication only; the other is similar. Suppose Σ is satisfiable in ML_p; say M, i, a sat Σ, where $M = (M_\sigma, m)_{\sigma \in P}$ is a standard model of ML_p based on D and I, $i \in I$ and $a \in As(M)$. Let $(M'_\alpha)_{\alpha \in T}$ be the standard frame for IL based on D and I, and define canonical one-to-one mappings Φ_σ from M_σ onto $M'_{\alpha[\sigma]}$, as follows:

(i) Φ_e is the identity mapping on $M_e = D = M'_e$,

(ii) For $\sigma = (\sigma_0,\ldots,\sigma_{n-1})$, $F \in M_\sigma = P(M_{\sigma_0} \times \ldots \times M_{\sigma_{n-1}})^I$, and $X'_k \in M'_{[\sigma_k]}$ ($k < n$), we put $\Phi_\sigma(F)(i)(X'_0) \ldots (X'_{n-1}) = 1$ just in case $(X_0,\ldots,X_{n-1}) \in F(i)$, where $X_k = \Phi_{\sigma_k}^{-1}(X'_k)$ ($k < n$).

We define a meaning function m' over the frame $(M'_\alpha)_{\alpha \in T}$ by letting $m'(\underline{c})(i) = \Phi_\sigma[m(c_\sigma)]$ for all $i \in I$, whenever c_σ is a constant of ML_p, and letting $m'(d_\alpha)$ be an arbitrary element of M'^I_α for constants d_α of IL which are not of the form $\underline{c}$. The system $M' = (M'_\alpha, m')_{\alpha \in T}$ is a standard model of IL, and one easily proves by induction:

LEMMA 13.1.1. Let A be a formula of ML_p. Suppose that $i \in I$, $a \in As(M)$, $a' \in As(M')$, and

(1) $a'(\underline{x}) = \Phi_\sigma[a(x_\sigma)]$

for every variable x_σ of ML_p. Then M, i, a sat A if and only if M', i, a' sat $\underline{A}$.

Since M, i, a sat Σ by assumption, if we choose $a' \in As(M')$ in such a way that (1) holds then the lemma yields M', i, a' sat $\underline{\Sigma}$. Since we clearly have M' sat Δ_Σ in view of the definition of m', we see that $\underline{\Sigma} \cup \Delta_\Sigma$ is satisfiable in IL, which completes the proof of the implication from left to right in Theorem 13.1.

COROLLARY 13.2. Let $\Sigma = \Gamma \cup \{A\}$ be a set of formulas of ML_P, and let $\Delta = \Delta_\Sigma$. Then $\Gamma \models A$ in ML_P if and only if $\underline{\Gamma} \cup \Delta \models \underline{A}$ in IL.

COROLLARY 13.3. Let A be a formula of ML_P and let δ_A be the conjunction of the formulas in $\Delta_{\{A\}}$. Then $\models A$ in ML_P if and only if $\models [\delta_A \to \underline{A}]$ in IL.

Turning now to the generalized semantics, we have the following:

THEOREM 13.4. Let Σ be a set of formulas of ML_P. If $\underline{\Sigma} \cup \Delta_\Sigma$ is g-satisfiable in IL, then Σ is g-satisfiable in ML_P+C+EC.

COROLLARY 13.5. Let $\Sigma = \Gamma \cup \{A\}$ be a set of formulas of ML_P, and let $\Delta = \Delta_\Sigma$. Then $\Gamma \models A$ in ML_P+C+EC implies $\underline{\Gamma} \cup \Delta \models \underline{A}$ in IL.

COROLLARY 13.6. Let A and δ_A satisfy the hypothesis of Corollary 13.3. Then $\models A$ in ML_P+C+EC implies $\models [\delta_A \to \underline{A}]$ in IL.

Proof of 13.4: Suppose M', i, a' satisfy $\underline{\Sigma} \cup \Delta_\Sigma$, where M' = $(M'_\alpha, m')_{\alpha \in T}$ is a g-model of IL based on D and I. Then M' satisfies each formula

(1) $\quad \exists x_{\alpha[\sigma]} [\underline{c} \equiv x]$,

where c_σ occurs in Σ, and in fact we can assume that M' satisfies (1) for all constants c_σ of ML_P, by redefining $m'(\underline{c})$ if necessary when c does not occur in Σ. We simultaneously define, by recursion on $\sigma \in P$, a set M_σ and a one-to-one mapping Φ_σ from M_σ onto $M'_{\alpha[\sigma]}$: We first put $M_e = D = M'_e$ and let Φ_e be the identity mapping on D. Next, we assume that $\sigma = (\sigma_0, \ldots, \sigma_{n-1})$ and M_{σ_k}, Φ_{σ_k} are defined for $k < n$. Given $F' \in M'_{\alpha[\sigma]}$, let $F \in P(M_{\sigma_0} \times \ldots \times M_{\sigma_{n-1}})^I$ be defined by the condition that $(X_0, \ldots, X_{n-1}) \in F(i)$ iff $F'(i)(X'_0) \ldots (X'_{n-1}) = 1$, where $X'_k = \Phi_{\sigma_k}(X_k)$ ($k < n$). It is easily checked that the mapping of F' to F is one-to-one on $M'_{\alpha[\sigma]}$; we let M_σ be its range and Φ_σ its inverse.

Clearly the family $(M_\sigma)_{\sigma \in P}$ defined in this way is a frame for ML_P based on D and I. We define a meaning function m by putting $m(c_\sigma) = \Phi_\sigma^{-1}[m'(\underline{c})(i)]$, which is independent of $i \in I$ by virtue of (1). The system $M = (M_\sigma, m)_{\sigma \in P}$ is a g-model of ML_P, and one verifies by induction that Lemma 13.1.1 holds in the present situation in exactly the form given earlier. Thus, since M', i, a' were assumed to satisfy $\underline{\Sigma}$, we must have also M, i, a sat Σ , where a is chosen to satisfy condition (1) of Lemma 13.1.1. It therefore remains only to show that M is a g-model of ML_p+C+EC , i.e., that the schemata of comprehension and extensional comprehension hold in M . Let

$$C^{\sigma,A} : \quad \exists f_\sigma \,\Box\, \forall x^0 \ldots \forall x^{n-1} \,[\, f\, x^0 \ldots x^{n-1} \leftrightarrow A\,]$$

be an instance of the comprehension schema, where $\sigma = (\sigma_0, \ldots, \sigma_{n-1})$, x^k is of type σ_k for $k < n$, and f_σ is the first variable of ML_p of type σ which is not free in the formula A . By Lemma 13.1.1 it suffices to show that its translate $\underline{C}^{\sigma,A}$ is true in M' ; in fact, we show the somewhat stronger:

LEMMA 13.4.1. Let α , α_0 , ... , $\alpha_{n-1} \in T$ and suppose that $\alpha = (s, (\alpha_0, (\ldots (\alpha_{n-1}, t) \ldots)))$. Let A be any formula of IL. Then the formula

$$\exists f_\alpha \,\Box\, \forall x^0 \ldots \forall x^{n-1} \,[\, {}^\vee f\, x^0 \ldots x^{n-1} \leftrightarrow A\,]$$

is provable in IL, where x^k is of type α_k for $k < n$ and f_α is the first variable of type α which does not occur free in A .

Proof: Let $M = (M_\alpha, m)_{\alpha \in T}$ be a g-model of IL, $a \in As(M)$. Putting $F = V_a^M(\, {}^\wedge \lambda x^0 \ldots \lambda x^{n-1} A\,) \in M_\alpha$, it is easily checked that

$$M;\, a, F \text{ sat } \Box\, \forall x^0 \ldots \forall x^{n-1} \,[\, {}^\vee f\, x^0 \ldots x^{n-1} \leftrightarrow A\,] \,.$$

In a similar way we verify that every instance $EC^{\sigma,A}$ of extensional comprehension is true in M by showing that the translate $\underline{EC}^{\sigma,A}$ is true in M' , and this follows from:

LEMMA 13.4.2. Under the hypotheses of Lemma 13.4.1, the formula

$$\Box\, \exists f_\alpha \,[\, \forall x^0 \ldots \forall x^{n-1} \,[\, \Box\, {}^\vee f\, x^0 \ldots x^{n-1} \vee \Box \sim {}^\vee f\, x^0 \ldots x^{n-1}\,]$$
$$\wedge\, \forall x^0 \ldots \forall x^{n-1} \,[\, {}^\vee f\, x^0 \ldots x^{n-1} \leftrightarrow A\,]\,]$$

is a theorem of IL.

Proof: Let $M = (M_\alpha, m)_{\alpha \in T}$ be a g-model of IL based on D and I , and suppose $i \in I$, $a \in As(M)$. We put $G = V_{i,a}^M(\, \lambda x^0 \ldots \lambda x^{n-1} A\,) \in M_\beta$, where $\beta = (\alpha_0,(\alpha_1,(\ldots(\alpha_{n-1},t)\ldots)))$, and let $F = V_G^M(\hat{g}_\beta) \in M_\alpha$. One verifies that M; i; F satisfy the formula

$$\forall x^0 \ldots \forall x^{n-1} [\Box \check{} f\, x^0 \ldots x^{n-1} \lor \Box \sim \check{} f\, x^0 \ldots x^{n-1}]$$
$$\land \forall x^0 \ldots \forall x^{n-1} [\check{} f\, x^0 \ldots x^{n-1} \leftrightarrow A],$$

which yields the desired result. This completes the proof of Theorem 13.4.

We remark that Corollary 13.6 can be given a direct syntactic proof. One shows that the set of formulas A of ML_p having the property that $[\underline{\delta_A} \to \underline{A}]$ is a theorem of IL contains the axioms of ML_p+C+EC and is closed under the inference rules of that theory.

<u>Interpretability of IL in ML_p</u>. We now outline a similar interpretation of IL in ML_p, omitting detailed proofs. First, we make correspond to each $\alpha \in T$ a type $\sigma[\alpha] \in P$:

(i) $\sigma[e] = e$,

(ii) $\sigma[t] = \phi$,

(iii) $\sigma[\alpha\beta] = (\sigma[\alpha],\sigma[\beta])$,

(iv) $\sigma[s\alpha] = (\sigma[\alpha])$.

For each $\alpha \in T$ and each variable $v_{\sigma[\alpha]}$ of ML_p we define a formula

$$T^\alpha(v) \quad (v \text{ \underline{is of type}} \quad \alpha)$$

of ML_p containing exactly the variable v free, as follows:

(i) $T^e(x_e)$ is $[x \equiv x]$,

(ii) $T^t(p_\phi)$ is $[\Box p \lor \Box \sim p]$,

(iii) $T^{\alpha\beta}(f_{\sigma[\alpha\beta]})$ is $Rn(f) \land \forall x_{\sigma[\alpha]} \forall y_{\sigma[\beta]} [fxy \to T^\alpha(x) \land T^\beta(y)] \land$
       $\forall x_{\sigma[\alpha]} [T^\alpha(x) \to \exists!!y_{\sigma[\beta]} \, fxy]$,

(iv) $T^{s\alpha}(f_{\sigma[s\alpha]})$ is $\Box \forall x_{\sigma[\alpha]} [fx \to T^\alpha(x)] \land \Box \exists!!x_{\sigma[\alpha]} \, fx$.

To each proper symbol s of IL we make correspond a symbol $\bar{s}$ of ML_p:

(i)   If s is x_α^n then $\bar{s}$ is $x_{\sigma[\alpha]}^n$,

(ii) If s is c_α^n then $\bar{s}$ is $c_{(\sigma[\alpha])}^n$.

Let A_α be a term of IL, x_σ a variable of ML_p. We say that x is <u>open for</u> A if x is distinct from $\bar{v}$ for every variable v which occurs free in A. For each term A_α of IL and each variable $x_{\sigma[\alpha]}$ of ML_p which is open for A_α, we define a formula

$$Co^A(x) \quad (\ x \ \underline{codes} \ A \ )$$

of ML_p whose free variables are x together with the variables $\bar{v}$ where v is free in A. The definition is by recursion on A_α:

(i) A_α is v_α, $x_{\sigma[\alpha]}$ open for A_α (i.e., x distinct from $\bar{v}$). Then $Co^A(x)$ is $[\ \bar{v} \equiv x\ ]$.

(ii) A_α is c_α, $x_{\sigma[\alpha]}$ arbitrary. Then $Co^A(x)$ is $\bar{c}x$.

(iii) A_α is $[B_{\beta\alpha} C_\beta]$, $y_{\sigma[\alpha]}$ open for A_α. Let f, x be the first variables of ML_p of types $(\sigma[\beta], \sigma[\alpha])$, $\sigma[\beta]$ respectively, which are distinct from y and open for A. Then $Co^A(y)$ is

$$\exists f\ \exists x\ [\ Co^B(f) \wedge Co^C(x) \wedge fxy\ ]\ .$$

(iv) A_α is $\lambda x_\beta\ B_\gamma$, $f_{\sigma[\alpha]}$ open for A_α. Let y be the first variable of type $\sigma[\gamma]$ which is distinct from $\bar{x}$ and open for B. Then $Co^A(f)$ is

$$Rn(f) \wedge \forall \bar{x}\ \forall y\ [\ f\bar{x}y \leftrightarrow T^\beta(\bar{x}) \wedge Co^B(y)\ ]\ .$$

(v) A_α is $[B_\beta \equiv C_\beta]$, p_ϕ open for A_α. Let x, y be the first distinct variables of type $\sigma[\beta]$ which are distinct from p and open for A. Then $Co^A(p)$ is

$$\exists x\ \exists y\ [\ Co^B(x) \wedge Co^C(y) \wedge \Box\ [\ p \leftrightarrow x \equiv y\ ]\ ]\ .$$

(vi) A_α is $\hat{\ }B_\beta$, $f_{\sigma[\alpha]}$ open for A_α. Let x be the first variable of type $\sigma[\beta]$ which is open for B. Then $Co^A(f)$ is

$$\Box\ \forall x\ [\ fx \leftrightarrow Co^B(x)\ ]\ .$$

(vii) A_α is $\check{\,}B_{s\alpha}$, $x_{\sigma[\alpha]}$ open for A_α . Let f be the first variable of type $(\sigma[\alpha])$ which is open for B . Then $Co^A(x)$ is

$$\exists f \,[\, Co^B(f) \wedge fx \,] \,.$$

For each formula A of IL we define its translate $\overline{A}$ in ML_p to be the formula $\exists p_\phi \,[\, Co^A(p) \wedge \Box p \,]$, where p is the first variable of type ϕ which is open for A . Given a set Σ of formulas of IL, we let $\overline{\Sigma}$ denote the set of all formulas $\overline{A}$ for $A \in \Sigma$. Also, we let Δ^Σ denote the set of all formulas (1) $T^{s\alpha}(\,\overline{c}\,)$, where c_α occurs in Σ , together with all formulas (2) $T^\alpha(\,\overline{x}\,)$, where x_α occurs free in Σ .

THEOREM 13.7. Let Σ be a set of formulas of IL. Then Σ is satisfiable in IL if and only if $\overline{\Sigma} \cup \Delta^\Sigma$ is satisfiable in ML_p.

We omit the proof.

COROLLARY 13.8. Let $\Sigma = \Gamma \cup \{A\}$ be a set of formulas of IL, and let $\Delta = \Delta^\Sigma$. Then $\Gamma \models A$ in IL if and only if $\overline{\Gamma} \cup \Delta \models \overline{A}$ in ML_p.

COROLLARY 13.9. Let A be a formula of IL, and let δ^A be the conjunction of the formulas in $\Delta^{\{A\}}$. Then $\models A$ in IL if and only if $\models [\, \delta^A \to \overline{A} \,]$ in ML_p.

For the generalized semantics, we state without proof the following analogue of Theorem 13.4:

THEOREM 13.10. Let Σ be a set of formulas of IL. If $\overline{\Sigma} \cup \Delta^\Sigma$ is g-satisfiable in ML_p+C+EC , then Σ is g-satisfiable in IL+D .

COROLLARY 13.11. Let $\Sigma = \Gamma \cup \{A\}$ be a set of formulas of IL, and let $\Delta = \Delta^\Sigma$. Then $\Gamma \models A$ in IL+D implies $\overline{\Gamma} \cup \Delta \models \overline{A}$ in ML_p+C+EC .

COROLLARY 13.12. Let A and δ^A satisfy the hypothesis of Corollary 13.9. Then $\models A$ in IL+D implies $\models [\, \delta^A \to \overline{A} \,]$ in ML_p+C+EC .

We remark that Corollary 13.12, like Corollary 13.6, can be proved directly without using generalized completeness. Combining Corollaries 13.6 and 13.12, we see that each of the theories IL+D , ML_p+C+EC is relatively interpretable in the other, in the following sense:

(i) $\vdash A$ in ML_p+C+EC implies $\vdash [\, \delta_A \to \underline{A}\,]$ in IL+D ,

(ii) $\vdash B$ in IL+D implies $\vdash [\, \delta^B \to \overline{B}\,]$ in ML_p+C+EC .

We state without proof:

THEOREM 13.13. Let A be a formula of ML_p, and let B be the formula $[\, \delta_A \to \underline{A}\,]$ of IL. Then $\vdash [\, \delta^B \to \overline{B}\,]$ in ML_p+C+EC implies $\vdash A$ in ML_p+C+EC . Similarly, let B be a formula of IL, and let A be the formula $[\, \delta^B \to \overline{B}\,]$. Then $\vdash [\, \delta_A \to \underline{A}\,]$ in IL+D implies $\vdash B$ in IL+D.

COROLLARY 13.14. The implications (i) and (ii) above can be strengthened to equivalence.

We say that the theories IL+D and ML_p+C+EC are <u>strongly</u> relatively interpretable in each other, in view of Corollary 13.14. We remark here, again without proof, that the theories IL+D and Ty_2+D (see §8) are also equivalent in the same sense: each is strongly relatively interpretable in the other. In fact, the interpretation of IL+D in Ty_2+D was given in §8 (see Theorem 8.3). For the interpretation of Ty_2+D in IL+D , we represent quantification over indices (objects of type s) by quantification, in IL, over atomic propositions in (approximately) the sense of §11.

CHAPTER 4.   ALGEBRAIC SEMANTICS

§14. Boolean Models of ML_p

In this section we describe an alternative semantics for the logic ML_p of Chapter 3, which will enable us to answer various independence questions raised earlier. The models with which we now concern ourselves are <u>Boolean</u> models, in distinction to the standard and general models of §9. This is an adaptation to higher-order modal logic of the notion of a Boolean model of ordinary higher-order predicate logic presented in Scott [1966]. The new feature is the presence in the language of the necessity operator $\square$, which turns out to be quite useful for describing various properties of the underlying algebra.[1]

Given sets $X_0, \ldots, X_{n-1}$, we say that R is an $(X_0,\ldots,X_{n-1})$-<u>relation</u> if $R \in P(X_0 \times \ldots \times X_{n-1})$, and given a set I we say that F is an $(I;X_0,\ldots,X_{n-1})$-<u>predicate</u> if $F \in P(X_0 \times \ldots \times X_{n-1})^I$. In view of the canonical set-theoretic equivalence

$$P(X_0 \times \ldots \times X_{n-1}) \cong 2^{X_0 \times \ldots \times X_{n-1}},$$

we can identify the $(X_0,\ldots,X_{n-1})$-relations with mappings from the product $X_0 \times \ldots \times X_{n-1}$ into the set $2 = \{0,1\}$ whose elements represent the respective truth-values falsity and truth. Under this identification, an $(X_0,\ldots,X_{n-1})$-relation R assigns to each n-tuple $(a_0,\ldots,a_{n-1})$ a truth-value $R(a_0,\ldots,a_{n-1})$, either 0 or 1. If we now replace the set 2 by an arbitrary Boolean algebra B, we obtain the set

$$B^{X_0 \times \ldots \times X_{n-1}}$$

---

[1] The basic idea behind the present construction is thus due to Scott, whose earlier work motivates most of this chapter. The author is indebted to Scott, in particular, for providing the general outline of §17.

of all B-__valued__ $(X_0,\ldots,X_{n-1})$-__relations__. Following Scott, we here think of B as comprising a widened class of truth-values: The zero and unit elements 0, 1 of B represent falsity and truth, while other elements $P \in B$ represent specific "degrees of truth" somewhere between them. If R is a B-valued $(X_0,\ldots,X_{n-1})$-relation, then R assigns to each n-tuple $(a_0,\ldots,a_{n-1}) \in X_0 \times \ldots \times X_{n-1}$ a Boolean truth-value $R(a_0,\ldots,a_{n-1}) \in B$, which we regard as the degree of truth of the assertion that $a_0, \ldots, a_{n-1}$ stand in the relation R. The ordinary $(X_0,\ldots,X_{n-1})$-relations can be identified with those B-valued relations which only assume the values 0 and 1. Now, from the equivalences

$$P(X_0 \times \ldots \times X_{n-1})^I \cong \left(2^{X_0 \times \ldots \times X_{n-1}}\right)^I$$
$$\cong \left(2^I\right)^{X_0 \times \ldots \times X_{n-1}}$$
$$\cong P(I)^{X_0 \times \ldots \times X_{n-1}},$$

we see that we can, for all practical purposes, identify the set of all $(I;X_0,\ldots,X_{n-1})$-predicates with the set of all B-valued $(X_0,\ldots,X_{n-1})$-relations, where B is the algebra $P(I)$ of all subsets of I. This suggests that the standard semantics for ML_p, which is based on domains

$$M_{(\sigma_0,\ldots,\sigma_{n-1})} = P(M_{\sigma_0} \times \ldots \times M_{\sigma_{n-1}})^I$$

of predicates, might be replaced by a Boolean semantics based on domains

$$M_{(\sigma_0,\ldots,\sigma_{n-1})} = B^{M_{\sigma_0} \times \ldots \times M_{\sigma_{n-1}}}$$

of B-valued relations. Of course, if B is the algebra $P(I)$ for some set I then this only amounts to a reformulation of the standard semantics for ML_p; we shall be interested, therefore, in the more general case.

Suppose that $B = (B, +, \cdot, -, 0, 1)$ is a complete Boolean algebra[2] and D is a non-empty set. The B-__valued Boolean frame for__ ML_p __based on__ D is the family $(M_\sigma)_{\sigma \in P}$ of sets, where:

---

[2] For the definition and basic properties of Boolean algebras, see Sikorski [1969] or Halmos [1963]. The hypothesis of completeness of the algebra will be necessary for the definition of the Boolean value of a formula; as Theorem 15.13 shows, this restriction cannot be dropped.

(i) $M_e = D$,

(ii) For $\sigma = (\sigma_0, \ldots, \sigma_{n-1})$, $M_\sigma = B^{M_{\sigma_0} \times \ldots \times M_{\sigma_{n-1}}}$.

A <u>B-valued Boolean model (b-model)</u> of ML_p based on D is a system $M = (M_\sigma, m)_{\sigma \in P}$ such that:

(i) $(M_\sigma)_{\sigma \in P}$ is the B-valued Boolean frame for ML_p based on D,

(ii) m (the meaning function) is a mapping which assigns to each constant c_σ an element of M_σ.

Let As(M) consist of all <u>assignments over</u> M; i.e., all functions a on the set of variables such that $a(x_\sigma) \in M_\sigma$ for each variable x_σ. Given $a \in As(M)$, let $\bar{a}$ extend a to the set of constants by the rule that $\bar{a}(c_\sigma) = m(c_\sigma) \in M_\sigma$. For each formula A of ML_p and each $a \in As(M)$ we define a <u>Boolean value</u> $\|A\|_a^M \in B$, as follows (we suppress the superscript 'M'):

(1) $\|s_\sigma s^0 \ldots s^{n-1}\|_a = \bar{a}(s_\sigma)(\bar{a}(s^0), \ldots, \bar{a}(s^{n-1}))$,

(2) $\|s_e \equiv s'_e\|_a = 1$ if $\bar{a}(s) = \bar{a}(s')$, 0 otherwise,

(3) $\|\sim A\|_a = -\|A\|_a$ (Boolean complement),

(4) $\|A \to B\|_a = [\|A\|_a \Rightarrow \|B\|_a] = -\|A\|_a + \|B\|_a$,

(5) $\|\forall x_\sigma A\|_a = \Pi_{X \in M_\sigma} \|A\|_{a,X}$ (Boolean infimum), where a,X represents the assignment $a(x/X)$, as usual,

(6) $\|\Box A\|_a = N\|A\|_a$, where N is the operator in B defined by: $NP = 1$ if $P = 1$, $NP = 0$ otherwise.

It is easily verified that $\|A\|_a$ depends only on the values of a for variables occurring free in A, so that, e.g., if $A(x_\sigma, y_\tau)$ involves only the distinct free variables x_σ, y_τ then we can write $\|A(x,y)\|_{X,Y}$ where $X \in M_\sigma$ and $Y \in M_\tau$, and if A is a closed formula we can write simply $\|A\|$. If A is modally closed then $\|A\|_a$ is either 0 or 1. A formula A is <u>true</u> in a b-model M if $\|A\|_a = 1$ for every assignment a, and A is <u>b-valid</u> in ML_p if it is true in every b-model.

We show now that the standard semantics for ML_P can in fact be viewed as a special case of the Boolean semantics, as suggested earlier. Let D and I be non-empty sets, let $(M_\sigma)_{\sigma \in P}$ be the standard frame for ML_P based on D and I, and let $(M_\sigma^*)_{\sigma \in P}$ be the $P(I)$-valued Boolean frame for ML_P based on D. We define canonical one-to-one mappings Φ_σ from M_σ onto M_σ^* as follows: We first let Φ_e be the identity mapping on $M_e = D = M_e^*$. For $\sigma = (\sigma_0, \ldots, \sigma_{n-1})$ we assume that Φ_{σ_k} is already defined for $k < n$. For each $F \in M_\sigma = P(M_{\sigma_0} \times \ldots \times M_{\sigma_{n-1}})^I$, we define its image

$$\Phi_\sigma(F) \in M_\sigma^* = P(I)^{M_{\sigma_0}^* \times \ldots \times M_{\sigma_{n-1}}^*}$$

by the condition that $i \in \Phi_\sigma(F)(X_0^*, \ldots, X_{n-1}^*)$ iff $(X_0, \ldots, X_{n-1}) \in F(i)$, where $X_k = \Phi_{\sigma_k}^{-1}(X_k^*)$ $(k < n)$.

THEOREM 14.1. Suppose $M = (M_\sigma, m)_{\sigma \in P}$ is a standard model of ML_P based on D and I, $M^* = (M_\sigma^*, m^*)_{\sigma \in P}$ is a $P(I)$-valued b-model of ML_P based on D, and for each constant c_σ we have $m^*(c_\sigma) = \Phi_\sigma[m(c_\sigma)]$. Given $a \in As(M)$ define $a^* \in As(M^*)$ by putting $a^*(x_\sigma) = \Phi_\sigma[a(x_\sigma)]$ for each variable x_σ. Then for every formula A and every $a \in As(M)$:

$$\| A \|_{a^*}^{M^*} = \{ i \in I \mid M, i, a \text{ sat } A \} .$$

Proof: Straightforward, by induction on A.

COROLLARY 14.2. Let A be a formula of ML_P. If A is b-valid in ML_P then A is valid in ML_P.

Before stating the next theorem on b-validity (Theorem 14.3), we need several additional lemmas. The first is a counterpart for the Boolean semantics of Lemma 9.1.1, and is easily proved by induction:

LEMMA 14.3.1. Let M be a b-model of ML_P and suppose the symbol s_σ is free for the variable x_σ in the formula $A(x)$. Then for every assignment a over M we have $\| A(s) \|_a = \| A(x) \|_{a,X}$, where $X = \bar{a}(s)$.

We recall from §9 that $[s_\sigma \equiv s'_\sigma]$ abbreviates the formula $\Box [s = s']$ for any type $\sigma \in P$, and if $\sigma = (\sigma_0, \ldots, \sigma_{n-1})$ then this in turn abbreviates

$$\Box \forall x^0_{\sigma_0} \ldots \forall x^{n-1}_{\sigma_{n-1}} [\ s\ x^0 \ldots x^{n-1} \leftrightarrow s'x^0 \ldots x^{n-1}\ ] .$$

From this we immediately obtain:

LEMMA 14.3.2. Let M be a B-valued b-model of ML_p, where B is a complete Boolean algebra. Let s_σ, s'_σ be any symbols of type σ. Then for every $a \in As(M)$, $\| s \equiv s' \|_a = 1$ if $\bar{a}(s) = \bar{a}(s')$, 0 otherwise.

We can now prove:

THEOREM 14.3. Every theorem of $ML_p + C$ is b-valid in ML_p.

Proof: We refer to the axioms and inference rules of the theory $ML_p + C$ set out in §9, i.e., axioms AS1 through AS9 (pp. 73-74), all instances of the comprehension schema (p. 77), and inference rules R1 - R3 (p. 74). It is clear that the rules preserve b-validity, so it suffices to show that every axiom is b-valid. For axioms AS1, AS2, A4 and A5 this follows immediately from the definition of Boolean value and elementary Boolean laws. Lemma 14.3.1 shows any instance

$$\forall x_\sigma\ A(x) \rightarrow A(s_\sigma)$$

of AS3 to be b-valid, since

$$\| \forall x_\sigma\ A(x) \|_a \leq \| A(x) \|_{a,X} = \| A(s_\sigma) \|_a ,$$

where $X = \bar{a}(s_\sigma)$. Similarly, given an instance of AS6, say

(1) $\qquad s_\sigma \equiv s'_\sigma \rightarrow [\ A(s) \rightarrow A(s')\ ] ,$

we show that it is true in any b-model M: Suppose $a \in As(M)$; then by Lemma 14.3.2., $\bar{a}(s) \neq \bar{a}(s')$ implies $\| s \equiv s' \|_a = 0$, which gives (1) the Boolean value 1. On the other hand, by Lemma 14.3.1 we have that $\bar{a}(s) = \bar{a}(s')$ implies $\| A(s) \|_a = \| A(x_\sigma) \|_{a,X} = \| A(s') \|_a$, where $X = \bar{a}(s) = \bar{a}(s')$, so that $\| A(s) \rightarrow A(s') \|_a = 1$, and therefore again (1) has Boolean value 1.

The modal axioms AS7, AS8 and AS9 are readily verified to be b-valid using the definition of Boolean value. For AS9, for instance, it suffices to show that if $B = (B, +, \cdot, -, 0, 1)$ is a complete Boolean algebra and N is the operator on B which interprets necessity -- i.e., $NP = 1$ if $P = 1$, $NP = 0$ otherwise -- then we have $-NP \leq N - NP$ for all $P \in B$. This is immediate.

Finally, suppose that

$$C^{\sigma, A} : \exists f_\sigma \Box \forall x^0 \ldots \forall x^{n-1} [f x^0 \ldots x^{n-1} \leftrightarrow A]$$

is an instance of the comprehension schema, where $\sigma = (\sigma_0, \ldots, \sigma_{n-1})$, x^k is of type σ_k for $k < n$, and f_σ is the first variable of type σ which does not occur free in A. Let $M = (M_\sigma, m)_{\sigma \in P}$ be a B-valued b-model of ML_p, $a \in As(M)$. Define $F \in B^{M_{\sigma_0} \times \ldots \times M_{\sigma_{n-1}}}$ by the condition $F(X_0, \ldots, X_{n-1}) = \| A \|_{a, X_0, \ldots, X_{n-1}}$. Then clearly we have

$$\| \forall x^0 \ldots \forall x^{n-1} [f x^0 \ldots x^{n-1} \leftrightarrow A] \|_{a, F} = 1 ,$$

so that $\| C^{\sigma, A} \|_a = 1$. This completes the proof.

By Corollary 14.2, Theorem 14.3 and generalized completeness for $ML_p + C$ we have:

COROLLARY 14.4. Let A be a formula of ML_p. Of the conditions

(i) A is g-valid in $ML_p + C$,

(ii) A is b-valid in ML_p,

(iii) A is valid in ML_p,

we have (i) implies (ii) implies (iii).

We shall see presently that these implications cannot be reversed, and also that condition (ii) is actually closer to (iii) than it is to (i).

M-Formulas. As in §12, we can simplify the present semantics somewhat by using constants in place of free variables. Suppose that $M = (M_\sigma, m)_{\sigma \in P}$ is a b-model of ML_p. For each $X \in M_\sigma$ we add to the vocabulary a new constant of type σ to denote X, and as earlier we simplify the discussion by agreeing to let X act as a name for itself, extending the meaning function m by putting $m(X) = X$ for each $X \in M_\sigma$, $\sigma \in P$. A formula of this extended language is called an M-<u>formula</u>, as before, and an M-<u>sentence</u> if it has no free variables. For M-sentences the Boolean value $\| A \|$ can be defined by recursion on the length of A; in particular, we have

$$\| c_\sigma c^0 \ldots c^{n-1} \| = m(c)(m(c^0), \ldots, m(c^{n-1}))$$

for any (extended) constants c, c^0, ..., c^{n-1}, and at the quantifier clause:

$$\| \forall x_\sigma A(x) \| = \prod_{X \in M_\sigma} \| A(X) \|.$$

It is easy to establish that $\| A(x^0, \ldots, x^{n-1}) \|_a = \| A(X_0, \ldots, X_{n-1}) \|$, where $A(x^0, \ldots, x^{n-1})$ is a formula of ML_p with distinct free variables x^k of type σ_k ($k < n$), and $a(x^k) = X_k$. We shall make free use of M-formulas in subsequent sections, concluding now with the following observations: In a B-valued Boolean model $M = (M_\sigma, m)_{\sigma \in P}$, the set $M_\phi = B^{\{\phi\}}$ can be identified with the algebra B itself,[3] so that a <u>proposition of</u> M is just an element $P \in B$, and considering P as an M-sentence we see that $\| P \| = P$. Similarly, a <u>propositional operator of</u> M will be a mapping $F \in M_{(\phi)} = B^{M_\phi} = B^B$, hence an operator on B in the usual sense.

§15. Modal Independence Results

We remarked in §9 that the schema EC of extensional comprehension is independent of $ML_p + C$, and again in §11 that the axiom At of atomic propositions is independent of $ML_p + C$. Indeed, by Theorem 11.5 these independence results are equivalent.

LEMMA 15.1.1. Let B be a complete Boolean algebra, M a B-valued Boolean model. Then $\| At \| = 1$ if B is atomic, and 0 otherwise.

---

[3] Recall that, by our earlier convention (page 72) regarding the Cartesian product of zero sets, we have $M_{\sigma_0} \times \ldots \times M_{\sigma_{n-1}} = \{\phi\}$ when $n = 0$.

Proof: Since the formula

At : $\Box \exists p_\phi [p \wedge \forall q_\phi [q \rightarrow \Box [p \rightarrow q]]]$

is modally closed, the value $\| At \|$ is either 0 or 1. But the following conditions are equivalent:

(1) $\| At \| = 1$,

(2) $\Sigma_{P \in B} \| P \wedge \forall q_\phi [q \rightarrow \Box [P \rightarrow q]] \| = 1$,

(3) $\Sigma_{P \in B} [P \cdot \Pi_{Q \in B} [Q \Rightarrow N[P \Rightarrow Q]]] = 1$,

(4) $\Sigma_{P \in B} [P \cdot \Pi_{Q \in B, P \not\leq Q} -Q] = 1$,

(5) $\Sigma_{P \in B, 0 < P} \Pi_{Q \in B, 0 < P \cdot -Q} P \cdot -Q = 1$,

(6) $\Sigma_{P \in B, 0 < P} \Pi_{R \in B, 0 < R \leq P} R = 1$,

(7)   B is atomic.

The equivalence of (3) and (4) follows from the fact that $N[P \Rightarrow Q] = 1$ if $P \leq Q$ and 0 otherwise. For the equivalence of (6) and (7) it suffices to observe that $\Pi_{R \in B, 0 < R \leq P} R$ is equal to P if P is an atom of B, and 0 otherwise.

If we now take for B any complete non-atomic Boolean algebra, e.g., the algebra of regular open subsets of the real line,[1] and take M to be an arbitrary B-valued Boolean model, we will have $\| At \| = 0$ by Lemma 15.1.1, so that the formula At is not b-valid. By Theorem 14.3 we immediately conclude:

THEOREM 15.1.[2] The formula At is not provable in ML_p+C .

In view of Lemmas 11.5.2 and 9.4 we also have

THEOREM 15.2. The formula E^σ is not provable in ML_p+C for $\sigma \neq e$, $\overline{n}$ $(n \in \omega)$.

---

[1] See Sikorski [1969], p. 5.

[2] This result can also be formulated so as to assert the consistency of the theory MLp+C+ ~At relative to that of a sufficiently strong theory, e.g., higher-order number theory. The set-theoretic proof of Theorem 15.1 can be replaced in this way by a finitary relative consistency proof.

COROLLARY 15.3. The schema EC is not derivable in ML_p+C.

As remarked in §11, E^ϕ is provable in ML_p+C, but Theorem 15.2 does not resolve the status of the formulas $E^{\bar{n}}$ for $n > 0$; we now turn to this question. Recall that, for a type $\sigma = (\sigma_0,\ldots,\sigma_{n-1})$, E^σ is the formula $\Box \forall f_\sigma \exists g_\sigma [Rn(g) \wedge f \equiv g]$, where $Rn(g)$ abbreviates

$$\forall x^0 \ldots \forall x^{n-1} [\Box g x^0 \ldots x^{n-1} \vee \Box \sim g x^0 \ldots x^{n-1}]$$

and $[f \equiv g]$ abbreviates

$$\forall x^0 \ldots \forall x^{n-1} [f x^0 \ldots x^{n-1} \leftrightarrow g x^0 \ldots x^{n-1}].$$

We first need:

LEMMA 15.4.1. Let B be a complete Boolean algebra, $M = (M_\sigma, m)_{\sigma \in P}$ a B-valued Boolean model, and suppose $\sigma = (\sigma_0,\ldots,\sigma_{n-1})$. Then for any $F \in M_\sigma$, we have $\|Rn(F)\| = 1$ just in case F is an ordinary two-valued $(M_{\sigma_0}, \ldots, M_{\sigma_{n-1}})$-relation, and $\|Rn(F)\| = 0$ otherwise.

Proof: Straightforward.

A complete Boolean algebra B is λ-<u>distributive</u>, where λ is a given cardinal (initial ordinal), if the identity

$$\Pi_{\xi < \lambda} \Sigma_{\eta < \lambda} P_{\xi,\eta} = \Sigma_{\varphi:\lambda \to \lambda} \Pi_{\xi < \lambda} P_{\xi,\varphi(\xi)}$$

holds for every doubly indexed family $(P_{\xi,\eta})_{\xi,\eta<\lambda}$ of elements of B, or equivalently, if

$$\Sigma_{K \subseteq \lambda} [\Pi_{\xi \in K} P_\xi \cdot \Pi_{\xi \notin K} -P_\xi] = 1$$

for every family $(P_\xi)_{\xi<\lambda}$ of elements of B.[3] Every Boolean algebra is λ-distributive for $\lambda < \omega$, and a complete algebra B is atomic if and only if it is completely distributive, i.e., λ-distributive for every λ.[4]

LEMMA 15.4.2. Suppose $n \in \omega$, $0 < n$. Let B be a complete Boolean algebra, $M = (M_\sigma, m)_{\sigma \in P}$ a B-valued Boolean model based on a set D of

---

[3] Sikorski [1969], pp. 61-62.
[4] Ibid., p. 105.

cardinality λ. Then $\|E^{\bar{n}}\| = 1$ if B is λ-distributive, 0 otherwise.

Proof: The formula in question is modally closed, hence has Boolean value 0 or 1. Note that

$$M_{\bar{n}} = M_{(e,\ldots,e)} = B^X,$$

where $X = D \times \ldots \times D = D^n$. The following are equivalent:

(1) $\quad \|E^{\bar{n}}\| = 1$,

(2) $\quad \Pi_{F:X \to B} \|\exists g_{\bar{n}} [Rn(g) \wedge F \equiv g]\| = 1$,

(3) $\quad$ For every $F:X \to B$, $\Sigma_{G:X \to B} \|Rn(G) \wedge F \equiv G\| = 1$,

(4) $\quad$ For every $F:X \to B$, $\Sigma_{G:X \to 2} \|F \equiv G\| = 1$,

(5) $\quad$ For every $F:X \to B$, $\Sigma_{G:X \to 2} \Pi_{a \in X} \|Fa \leftrightarrow Ga\| = 1$,

(6) $\quad$ For every $F:X \to B$,

$$\Sigma_{G:X \to 2} [\Pi_{G(a) = 1} F(a) \cdot \Pi_{G(a) = 0} -F(a)] = 1,$$

(7) $\quad$ For every family $(P_\xi)_{\xi < \kappa}$, where $\kappa = \lambda^n = |X|$,

$$\Sigma_{K \subseteq \kappa} [\Pi_{\xi \in K} P_\xi \cdot \Pi_{\xi \notin K} -P_\xi] = 1,$$

(8) $\quad B$ is κ-distributive,

(9) $\quad B$ is λ-distributive.

The equivalence of (3) and (4) uses Lemma 15.4.1. The equivalence of (8) and (9) follows from the fact that $\lambda < \omega$ implies $\kappa = \lambda^n < \omega$ also, in which case both (8) and (9) hold, while $\lambda \geq \omega$ implies $\kappa = \lambda^n = \lambda$, since $0 < n$.

THEOREM 15.4. The formula $E^{\bar{n}}$ is not provable in ML_p+C for $n > 0$.

Proof: Let B be any non-atomic complete Boolean algebra. Then for some cardinal λ, B is not λ-distributive. Take for D any set of cardinality λ, and let M be any B-valued Boolean model based on D. Then by Lemma 15.4.2, we have $\|E^{\bar{n}}\| = 0$ in M, so that $E^{\bar{n}}$ is not b-valid and therefore, by Theorem 14.3, not provable in ML_p+C.

In contrast to Lemma 11.5.2, we also have the following result:

THEOREM 15.5. The formula $[E^{\bar{n}} \to At]$ is not provable in ML_p+C for any $n \in \omega$.

Proof: If $n = 0$ this follows from Theorem 15.1 and the fact that E^ϕ is provable in ML_p+C. For $n > 0$ it suffices to observe that $\| E^{\bar{n}} \| = 1$ and $\| At \| = 0$ in any B-valued Boolean model based on D, where B is chosen to be non-atomic and D is finite.

The conditions E^σ of extensional comprehension can be weakened; specifically, for a predicate type $\sigma = (\sigma_0, \ldots, \sigma_{n-1})$ we consider the formula

$$E_2^\sigma : \quad \forall f_\sigma \exists g_\sigma [\, Rn(g) \wedge \Diamond [f \equiv g] \,] \;.$$

This formula, which is modally closed, asserts that every predicate is co-extensional, at some index, with an ordinary relation. By generalized completeness we immediately have:

LEMMA 15.6.1. The formula $[E^\sigma \to E_2^\sigma]$ is provable in ML_p+C.

Thus, E_2^ϕ is provable in ML_p+C, and E_2^σ is valid in ML_p (i.e., with respect to the standard semantics) for every $\sigma \neq e$. We now show that the conditions E_2^σ are essentially weaker than the conditions E^σ.

LEMMA 15.6.2. Let B be a complete Boolean algebra. Then B is atomless if and only if there exists a set $L \subseteq B$ such that for every $K \subseteq L$:

$$\Pi_{P \in K} P \leq \Sigma_{P \in L-K} P \;.$$

Proof: First assume B is atomless. Let $L = B$ and suppose $K \subseteq L$. If $\Pi K = \Pi_{P \in K} P = 0$, then the condition holds, so we may assume that $0 < \Pi K$. Since B is atomless, there exist $P, Q \in B = L$ with $0 < P$, $0 < Q$ and $P + Q = \Pi K$. But then $P, Q \in L-K$, so that we have $\Pi_{P \in K} P = \Pi K = P + Q \leq \Sigma_{P \in L-K} P$. Conversely, assume the condition holds for some set $L \subseteq B$, and let Q be an atom of B. Define $K \subseteq L$ by putting $P \in K$ just in case $P \in L$ and $Q \leq P$. Since $Q \leq P$ for all $P \in K$ we have $Q \leq \Pi_{P \in K} P \leq \Sigma_{P \in L-K} P$, whence $Q \leq P$ for some $P \in L-K$, since Q is an atom. This is a contradiction, so B must in fact be atomless.

Using Lemma 15.6.2, we can show the following:

LEMMA 15.6.3. Suppose $\sigma \neq e, \phi$. Let B be a complete Boolean algebra, $M = (M_\sigma, m)_{\sigma \in P}$ a B-valued Boolean model based on a set D, where $|D| \geq |B|$. Then $\|E^\sigma\| = 1$ iff B is atomic, and $\|E_2^\sigma\| = 1$ iff B contains an atom.

Proof: Say $\sigma = (\sigma_0, \ldots, \sigma_n)$. Then $M_\sigma = B^X$, where X is the Cartesian product $M_{\sigma_0} \times \ldots \times M_{\sigma_n}$. It is easily seen that $\lambda \geq |B|$, where $\lambda = |X|$. As before, in the proof of Lemma 15.4.2, one verifies the fact that $\|E^\sigma\| = 1$ if and only if B is λ-distributive, which is equivalent since $\lambda \geq |B|$ to the condition that B is completely distributive, i.e., atomic. On the other hand, we have the following equivalences:

(1) $\|E_2^\sigma\| = 1$,

(2) $\Pi_{F:X \to B} \|\exists g_\sigma [Rn(g) \wedge \Diamond[F \equiv g]]\| = 1$,

(3)   For every $F:X \to B$, $\Sigma_{G:X \to 2} \|\Diamond[F \equiv G]\| = 1$,

(4)   For every $F:X \to B$, there is $G:X \to 2$ with $\|F \equiv G\| \neq 0$,

(5)   For every $F:X \to B$, there is $G:X \to 2$ such that

$$\Pi_{a \in X, G(a) = 1} F(a) \cdot \Pi_{a \in X, G(a) = 0} -F(a) \neq 0,$$

(6)   For every family $(P_\xi)_{\xi < \lambda}$ of elements of B, there exists a set $K \subseteq \lambda$ such that

$$\Pi_{\xi \in K} P_\xi \cdot \Pi_{\xi \notin K} -P_\xi \neq 0,$$

(7)   For every $L \subseteq B$ there exists a set $K \subseteq L$ such that

$$\Pi_{P \in K} P \nleq \Sigma_{P \in L-K} P,$$

(8) B contains an atom.

We use here Lemmas 15.4.1, 15.6.2, and the fact that $\lambda \geq |B|$. This completes the proof. Choosing for B an atomless complete Boolean algebra (e.g., the algebra of regular open subsets of the real line), we obtain:

THEOREM 15.6. For $\sigma \neq e, \phi$, E_2^σ is not provable in $ML_p + C$.

THEOREM 15.7. For $\sigma \neq e, \phi$, E^σ is not derivable in $ML_p + C$ from the set of formulas $\{E_2^\sigma \mid \sigma \neq e\}$.

Proof: Assume otherwise; then some formula

$$E_2^{\sigma_0} \to . \; E_2^{\sigma_1} \to . \; \ldots ; \to . \; E_2^{\sigma_{n-1}} \to E^{\sigma}$$

is provable in ML_p+C and hence b-valid in ML_p. Choose for B a complete Boolean algebra which is neither atomic nor atomless, e.g., a direct product $B_1 \times B_2$, where B_1 is atomic and B_2 is atomless. Let D be any set with $|D| \geq |B|$, and let M be any B-valued b-model based on D. By Lemma 15.6.3, the formula above has Boolean value 0 in M, a contradiction.

It was remarked in §12 that there exist g-models of ML_p+C which contain non-indicial S5-operators. We can now prove this, by the following argument: The question whether such operators exist is just the question whether the formula

(*) $\quad \forall f_{(\phi)} \; [\; A_2(f) \wedge R_2(f) \wedge A_3(f) \wedge A_6(f) \to$

$$\Box \exists p_\phi \forall q_\phi \; [\; fq \leftrightarrow \Box \; [p \to q] \;] \;] \; ,$$

which asserts that every S5-operator is indicial, is g-valid -- or equivalently, provable -- in ML_p+C.

THEOREM 15.8. The formula (*) is not b-valid in ML_p, and therefore is not provable in ML_p+C.

Proof: Let B be any non-atomic complete Boolean algebra, and let $M = (M_\sigma, m)_{\sigma \in P}$ be a B-valued Boolean model. If $F \in M_{(\phi)} = B^B$ is any propositional operator of M, and A is any M-sentence, then FA abbreviates the M-sentence $\exists p_\phi [\; \Box[p \leftrightarrow A] \wedge Fp \;]$, from which it follows that $\| FA \| = F(\| A \|)$. Now suppose that F is the identity operator, defined by $F(P) = P$ for all $P \in B$. Then we have $\| FA \| = \| A \|$ for every M-sentence A. From this it is easily verified that each of the M-sentences $A_2(F)$, $R_2(F)$, $A_3(F)$ and $A_6(F)$ has Boolean value 1 in M; i.e., F is an S5-operator of the Boolean model M. On the other hand, the M-sentence

(1) $\quad \Box \exists p_\phi \forall q_\phi \; [\; Fq \leftrightarrow \Box \; [p \to q] \;]$

asserting that F is indicial, is equivalent in M to the sentence

(2) $\Box \exists p_\phi \forall q_\phi [q \leftrightarrow \Box [p \rightarrow q]]$,

which is provably equivalent in ML_p+C to the axiom At of atomic propositions. Hence, by Lemma 15.1.1, formula (1) has Boolean value 0 in M, and therefore (*) also has value 0 in M.

The independence results presented in this section were all proved by means of the Boolean methods developed in §14. Whether the same results could be obtained without these methods has not been rigorously settled, although the author strongly believes that direct proofs using only the generalized semantics for ML_p are not possible. To complete the discussion of independence, however, we mention the following result, whose proof does not require Boolean semantics:

THEOREM 15.9. The comprehension schema C is not derivable in the theory ML_p+EC whose axioms are those of ML_p together with all instances $EC^{\sigma,A}$ of the extensional comprehension schema. In fact, the instance

$C^{\phi,fx}$: $\exists p_\phi \Box [p \leftrightarrow f_{(e)} x_e]$

is not provable in ML_p+EC.

Proof: Let D and I be arbitrary sets, with $|I| > 1$. Define a frame $(M_\sigma)_{\sigma \in P}$ for ML_p based on D and I by:

(i) $M_e = D$,
(ii) $M_\phi = \{ P \in 2^I \mid P \text{ constant} \} = \{P_0, P_1\}$,
(iii) For $\sigma = (\sigma_0, \ldots, \sigma_n)$, $n \geq 0$, $M_\sigma = P(M_{\sigma_0} \times \ldots \times M_{\sigma_n})^I$.

Let $m(c_\sigma) \in M_\sigma$ be arbitrary. It is obvious that every instance $EC^{\sigma,A}$ of extensional comprehension is true in M. But clearly, for an appropriate choice of $F \in M_{(e)} = P(D)^I$ and $X \in D$ we will not have M; F,X sat $C^{\phi,fx}$.

Validity and B-Validity. We have seen (Corollary 14.2) that every b-valid formula of ML_p is valid. The converse is not true, since by Lemma 15.1.1, the axiom At of atomic propositions is not b-valid. We proceed to show, however, that it is only At that stands between validity and b-validity.

THEOREM 15.10. Let A be a formula of ML_p. Then A is valid in ML_p if and only if $[At \to A]$ is b-valid in ML_p.

Proof: Suppose first that $[At \to A]$ is b-valid. Then it is valid, by Corollary 14.2, and since At is valid it follows that A is valid. Conversely, suppose A is valid, and let $M^* = (M_\sigma^*, m^*)_{\sigma \in P}$ be a B-valued Boolean model of ML_p, a^* an assignment over M^*. If B is non-atomic then $\| At \| = 0$ by Lemma 15.1.1, and therefore $\| At \to A \|_{a^*} = 1$. We can therefore assume that B is atomic. Since B is complete, it follows that B is isomorphic to a complete field of sets; in fact, to the algebra $P(I)$ of all subsets of the set I of atoms of B.[5] We can therefore identify B with $P(I)$. Now let $(M_\sigma)_{\sigma \in P}$ be the standard frame for ML_p based on D and I, where $D = M_e^*$, and let Φ_σ be the canonical one-to-one mapping from M_σ onto M_σ^* defined in §14. If we put $m(c_\sigma) = \Phi_\sigma^{-1}[m^*(c_\sigma)]$ for each constant c_σ and let $a(x_\sigma) = \Phi_\sigma^{-1}[a^*(x_\sigma)]$ for each variable x_σ, then by Theorem 14.1:

$$\| A \|_{a^*}^{M^*} = \{ i \in I \mid M, i, a \text{ sat } A \} = I ,$$

since A is valid in ML_p. Thus A has Boolean value 1 in M^*, under the assignment a^*, and the proof is complete.

COROLLARY 15.11. The b-valid formulas of ML_p are not axiomatizable. In fact, the set of Gödel numbers of b-valid formulas of ML_p is not definable by any formula of higher-order number theory.

Proof: It is known[6] that the set of valid formulas of L_p is not definable in higher-order number theory, and the same result clearly holds for the set of valid formulas of ML_p. This fact together with Theorem 15.10 shows that the set of b-valid formulas of ML_p is also undefinable, and therefore not recursively enumerable. The result can also be obtained by a direct interpretation of L_p in ML_p: By relativizing quantifiers to heredi-

---

[5] Sikorski [1969], p. 105.

[6] See Enderton [1972], p. 272 (The argument is based on Tarski's theorem on undefinability). L. Harrington informs us that the counterpart of Corollary 15.11 for ordinary (non-modal) higher-order logic is also true, although the proof lies much deeper.

tarily constant predicates, using the formula $Rn(f)$, we can effectively find for each formula A of L_p a formula A^* of ML_p such that A is valid in L_p just in case A^* is b-valid in ML_p. Since the mapping is recursive and hence arithmetically definable, the result follows.

Expressible Boolean Properties. We have seen that the property of atomicity of a complete Boolean algebra B can be expressed in ML_p by the formula At, in the sense of Lemma 15.1.1; viz., in any B-valued Boolean model we will have $\|At\| = 1$ if B is atomic, and 0 otherwise. There is, in fact, a wide class of Boolean properties expressible in ML_p in the same sense, i.e., by means of a closed and modally closed formula of ML_p. Specifically, we can express in ML_p any Boolean property expressible in the language of higher-order Boolean algebra, by which we understand a formalism containing primitives for the Boolean operations and relations, and variables over elements of the algebra, sets of elements, sets of sets, etc. To see this, we observe that variables p_ϕ of type ϕ range over elements of B in our Boolean semantics, and the Boolean relations $P \leq Q$ and $P = Q$ can be expressed by the modally closed formulas $\Box[p_\phi \to q_\phi]$ and $[p_\phi \equiv q_\phi]$, respectively. The Boolean operations $P + Q$, $P \cdot Q$, $-P$ can be expressed by the formulas $p \lor q$, $p \land q$, $\sim p$ respectively. The formula $\forall p_\phi A(p)$, where $A(p)$ is modally closed, expresses quantification over elements of the algebra, while $\forall f_{(\phi)} [Rn(f) \to A(f)]$ expresses quantification over subsets of the algebra, and similarly for higher orders. We therefore have:

THEOREM 15.12. For any sentence A of the language of higher-order Boolean algebra we can effectively find a closed and modally closed formula A^* of ML_p such that for any complete Boolean algebra B and any B-valued Boolean model M, $\|A^*\|^M = 1$ if and only if A is true in B.

Incomplete Boolean Models. We may attempt to relax the requirement of completeness on the algebra B in our definition of Boolean model. However, the next theorem shows that it is in fact not possible to do so. Let B be an arbitrary Boolean algebra, not necessarily complete, D a nonempty set. A B-valued Boolean model of ML_p based on D is a system $M = (M_\sigma, m)_{\sigma \in P}$ such that:

(i) $(M_\sigma)_{\sigma \in P}$ is the B-valued Boolean frame for ML_p based on D,

(ii) m assigns to each constant c_σ an element of M_σ,

(iii) There exists a function $\| \ \|^M$ which assigns, to each formula A of ML_p and $a \in As(M)$, a Boolean value $\| A \|_a^M \in B$, in such a way as to satisfy the recursive conditions (1) through (6) on page 108.

Clause (iii) should be interpreted as follows with respect to condition (5) on page 108: Given a formula $\forall x_\sigma A$ and an assignment a, the Boolean infimum $\Pi_{X \in M_\sigma} \| A \|_{a,X}^M$ exists in B and is equal to $\| \forall x_\sigma A \|_a^M$.

THEOREM 15.13. If there exists a B-valued Boolean model then B is complete.

Proof: Suppose $M = (M_\sigma, m)_{\sigma \in P}$ is a B-valued Boolean model, where B is an arbitrary Boolean algebra. Suppose $K \subseteq B$, and define an operator $F \in M_{(\phi)} = B^B$ by putting $F(P) = P$ if $P \in K$, $F(P) = 1$ otherwise. Then by clause (iii) we have $\| \forall p_\phi f_{(\phi)} p_\phi \|_F^M = \Pi_{P \in M_\phi} \| f_{(\phi)} p_\phi \|_{F,P}^M = \Pi_{P \in B} F(P) = \Pi_{P \in K} P$, so that the infimum ΠK of K belongs to B. Since K was arbitrary, B is complete.

In the next section, we consider a somewhat different generalization of the notion of a Boolean model.

§16. Topological Models of ML_p

Since the early 1940's, one branch of the study of propositional modal logics has centered around topological interpretations, or more generally, interpretations in such structures as topological Boolean algebras[1], which are naturally related to topological spaces. A <u>topological Boolean algebra</u> is a structure (B,N), where B is a Boolean algebra and N is an operator on B satisfying:

(i) $NP \leq P$,

(ii) $N[P \cdot Q] = NP \cdot NQ$,

---

[1] Also called closure algebras; see McKinsey [1941], McKinsey and Tarski [1944]. We here employ the dual notion, as in Rasiowa [1963].

(iii) $NNP = NP$,

(iv) $N1 = 1$.

The motivating example is a <u>topological field of sets</u> (B,N) , where X is a topological space with interior operator N , and B is a field of subsets of X closed under N . It can be shown[2] that every topological Boolean algebra is isomorphic to a topological field of sets.

In any topological algebra (B,N) we can define the usual topological notions: An element $P \in B$ is <u>open</u> if $NP = P$, or equivalently if $NQ = P$ for some $Q \in B$. The <u>closure</u> of an element P is defined to be $-N[-P]$, and P is <u>closed</u> if it is equal to its closure, or equivalently if $-P$ is open. The infimum $P_1 \cdot P_2 \cdot \ldots \cdot P_n$ of finitely many open elements is again open, and the supremum $\Sigma_{P \in K} P$ of any set K of open elements is open, if it exists in B . For any $P \in B$, NP equals the supremum $\Sigma_{Q \leq P, NQ = Q} Q$ of all open elements dominated by P .

It is easily verified that clause (ii) of the definition of topological algebra can be replaced by

(ii') $N[P \Rightarrow Q] \leq [NP \Rightarrow NQ]$,

so that a topological algebra (B,N) consists of a Boolean algebra B together with a Boolean S4-operator on B , in the obvious sense. It is this fact that provides the connection, remarked earlier, between topological algebras and modal logic. Accordingly, we say that a topological algebra (B,N) is an <u>S5-algebra</u> if it satisfies also

(v) $-NP \leq N[-NP]$

for all $P \in B$, or equivalently, if every closed element is also open (and hence conversely).[3] In such an algebra, the open elements form a Boolean subalgebra B^* of B , which is a complete subalgebra, in the sense that if $K \subseteq B^*$ and the supremum ΣK exists in B , then it belongs to B^* . Consequently:

---

[2] McKinsey and Tarski [1944].

[3] See McKinsey and Tarski [1948], p. 8.

LEMMA 16.1. Let (B,N) be a complete S5-algebra. Then the open elements form a complete subalgebra B^*, and for every $P \in B$ we have

(1) $NP = \Sigma_{Q \in B^*, Q \leq P} \; Q$.

Conversely, if B is any complete Boolean algebra, B^* is a complete subalgebra of B, and we define an operator $N \in B^B$ by condition (1), then (B,N) is a complete S5-algebra and B^* consists of its open elements.

Let (B,N) be a complete S5-algebra, D a non-empty set. We define the (B,N)-<u>valued topological frame for</u> ML_p <u>based on</u> D to be the family $(M_\sigma)_{\sigma \in P}$ of sets, where we simultaneously define the set M_σ and the Boolean value $\| X \equiv Y \|$ for $X, Y \in M_\sigma$ as follows:

(i) $M_e = D$; $\| X \equiv Y \| = 1$ if $X = Y$ in D, 0 otherwise.

(ii) For $\sigma = (\sigma_0,\ldots,\sigma_{n-1})$, M_σ consists of all mappings F from $M_{\sigma_0} \times \ldots \times M_{\sigma_{n-1}}$ into B such that for any sequences $X, Y \in M_{\sigma_0} \times \ldots \times M_{\sigma_{n-1}}$, $\Pi_{k < n} N \| X_k \equiv Y_k \| \leq [F(X) \Leftrightarrow F(Y)]$.[4] We let $\| F \equiv G \| = \Pi_{X \in M_{\sigma_0} \times \ldots \times M_{\sigma_{n-1}}} [F(X) \Leftrightarrow G(X)]$ for any $F, G \in M_\sigma$.

A (B,N)-<u>valued topological model (t-model) of</u> ML_p <u>based on</u> D is a system $M = (M_\sigma, m)_{\sigma \in P}$ such that:

(i) $(M_\sigma)_{\sigma \in P}$ is the (B,N)-valued topological frame based on D,

(ii) m assigns to each constant c_σ an element of M_σ .

Let $As(M)$ consist of all assignments over M, in the usual sense. For each formula A of ML_p and each $a \in As(M)$ we define $\| A \|_a^M \in B$ as in §14, except that condition (6) on page 108 is now replaced by:

(6') $\| \Box A \|_a = N \| A \|_a$, where N is the interior operator of the algebra (B,N) .

---

[4] In any Boolean algebra we write $[P \Leftrightarrow Q]$ for the Boolean combination $P \cdot Q + (-P) \cdot (-Q)$, where $P, Q \in B$. Condition (ii) is necessary in order to validate the equality axioms of MLp in the topological semantics. See the proof of Theorem 16.2.

A formula A is <u>true</u> in a t-model M if $\| A \|_a^M = 1$ for all $a \in As(M)$, and A is <u>t-valid</u> in ML_p if it is true in every t-model.[5]

THEOREM 16.2. Every theorem of ML_p+C is t-valid in ML_p.

<u>Proof</u>: Clearly the inference rules R1 - R3 of ML_p preserve t-validity. The proof that every axiom of ML_p is t-valid proceeds as in the proof of Theorem 14.3, except that one must verify that axiom schema AS6 can be replaced in ML_p by the following axioms:

A6.1 $x_e \equiv y_e \to y_e \equiv x_e$,

A6.2 $x_e \equiv y_e \to [\, y_e \equiv z_e \to x_e \equiv z_e \,]$,

A6.3 $x^0 \equiv y^0 \wedge \ldots \wedge x^{n-1} \equiv y^{n-1} \to [\, f\, x^0 \ldots x^{n-1} \to f\, y^0 \ldots y^{n-1} \,]$,

     where $x^0, \ldots, x^{n-1}, y^0, \ldots, y^{n-1}$ is the first sequence of distinct variables such that x^k and y^k are of type σ_k ($k < n$), and f is of type $\sigma = (\sigma_0, \ldots, \sigma_{n-1})$.

These axioms are easily seen to be t-valid -- in the case of A6.3 this follows from clause (ii) in the definition of a topological frame -- so that every theorem of ML_p is t-valid. It remains to verify the t-validity of each instance

$$C^{\sigma,A} : \quad \exists f_\sigma \,\Box\, \forall x^0 \ldots \forall x^{n-1} \,[\, f\, x^0 \ldots x^{n-1} \leftrightarrow A \,]$$

of the comprehension schema. Suppose that M is a (B,N)-valued topological model and $a \in As(M)$. As in the proof of Theorem 14.3 one defines a mapping F from $M_{\sigma_0} \times \ldots \times M_{\sigma_{n-1}}$ into B by letting $F(X_0, \ldots, X_{n-1}) = \| A \|_{a, X_0, \ldots, X_{n-1}}$. It must then be shown that $F \in M_\sigma$; but this follows from the fact that the formula

$$\forall x^0 \ldots \forall x^{n-1} \forall y^0 \ldots \forall y^{n-1} \,[\, x^0 \equiv y^0 \wedge \ldots \wedge x^{n-1} \equiv y^{n-1}$$
$$\to \,.\, A(x^0, \ldots, x^{n-1}) \leftrightarrow A(y^0, \ldots, y^{n-1}) \,]$$

is provable in ML_p and hence t-valid, where A is $A(x^0, \ldots, x^{n-1})$, the

---

[5] The idea of interpreting first-order modal logic in complete topological algebras is apparently due to Rasiowa [1951].

variables $y^0, \ldots, y^{n-1}$ are distinct from $x^0, \ldots, x^{n-1}$ and from each other and do not occur in A, and y^k is of the same type as x^k ($k < n$). We omit the details.

REMARKS. Suppose B is a complete Boolean algebra. The <u>trivial</u> operator $N \in B^B$ is defined by: $NP = 1$ if $P = 1$, $NP = 0$ otherwise. The pair (B,N) is an S5-algebra, and a (B,N)-valued topological model is just a B-valued Boolean model, as defined in §14 (This operator corresponds to the trivial topology). Thus, every b-model is also a t-model, and every t-valid formula is b-valid. If we define $N \in B^B$ by letting $NP = P$ for all $P \in B$ (the <u>discrete</u> operator), we obtain another S5-algebra (B,N), corresponding to the discrete topology. In any discrete topological model M, the formula $\Box A \leftrightarrow A$ is true for every formula A, so that the necessity operator $\Box$ is vacuous and hence M is essentially the same as one of Scott's Boolean models of ordinary higher-order logic. Thus, both Scott's Boolean models and those defined in §14 can be construed as topological models.

The notion of a topological model can be further generalized by permitting a smaller set of Boolean-valued relations at each level. Thus, if (B,N) is a complete S5-algebra, a <u>(B,N)-valued general topological frame</u> is defined as before except for the replacement of clause (ii) by:

(ii') For $\sigma = (\sigma_0, \ldots, \sigma_{n-1})$, M_σ is some non-empty set of mappings F from $M_{\sigma_0} \times \ldots \times M_{\sigma_{n-1}}$ into B such that for any sequences X, Y $\in M_{\sigma_0} \times \ldots \times M_{\sigma_{n-1}}$, $\prod_{k<n} N \| X_k \equiv Y_k \| \leq [F(X) \leftrightarrow F(Y)]$.

This leads to the notion of a <u>(B,N)-valued general topological model</u>. If N is the trivial operator on B, we speak instead of a <u>B-valued general Boolean model</u>. Every theorem of ML_p is true in every general topological model, but in general the comprehension axioms will not hold. Nevertheless, such general models prove useful, e.g., in showing the independence of the axiom of choice in ordinary higher-order logic.

<u>Homomorphisms</u>. Let (B,N) and (B',N') be topological algebras. A <u>homomorphism from</u> (B,N) <u>into</u> (B',N') is a mapping ϑ from B into B' such that for all $P, Q \in B$: (i) $\vartheta(P+Q) = \vartheta(P) + \vartheta(Q)$; (ii) $\vartheta(-P) = -\vartheta(P)$; (iii) $\vartheta(NP) = N'[\vartheta(P)]$. Here we use + and - ambiguously to

denote the operations of B' as well as those of B. A homomorphism ϑ is <u>complete</u> if it preserves infinite infima and suprema in B; i.e., if $\Sigma_{P \in K} \vartheta(P)$ exists in B' and is equal to $\vartheta(\Sigma_{P \in K} P)$ whenever $\Sigma_{P \in K} P$ exists in B. We shall be interested in complete homomorphisms ϑ on complete S5-algebras (B,N). In this case the range (B',N') of ϑ is also a complete S5-algebra, and all infima and suprema are preserved.[6]

Let $M = (M_\sigma, m)_{\sigma \in P}$ be a (B,N)-valued topological model based on D, where (B,N) is a complete S5-algebra, and suppose that ϑ is a complete homomorphism from (B,N) onto (B',N'). We define a (B',N')-valued topological model $M' = (M'_\sigma, m')_{\sigma \in P}$ also based on D, which we call the <u>image</u> of the model M under ϑ. We first let $(M'_\sigma)_{\sigma \in P}$ be the (B',N')-valued topological frame based on D, and define by recursion on σ mappings ϑ_σ from M_σ onto M'_σ satisfying the following condition:

(1)   For all $X, Y \in M_\sigma$, $\vartheta \| X \equiv Y \| = \| \vartheta_\sigma(X) \equiv \vartheta_\sigma(Y) \|$.

(Recall that the value $\| X \equiv Y \|$ is defined for $X, Y \in M_\sigma$ in any topological frame.) The definition of ϑ_σ is as follows: We first let ϑ_e be the identity mapping on $M_e = D = M'_e$. For $\sigma = (\sigma_0, \ldots, \sigma_{n-1})$, we assume that ϑ_{σ_k} maps M_{σ_k} onto M'_{σ_k} for $k < n$, in such a way that condition (1) holds for each σ_k. Given $F \in M_\sigma \subseteq B^{M_{\sigma_0} \times \ldots \times M_{\sigma_{n-1}}}$, we define $\vartheta_\sigma(F) \in B'^{M'_{\sigma_0} \times \ldots \times M'_{\sigma_{n-1}}}$ by: $\vartheta_\sigma(F)(X'_0, \ldots, X'_{n-1}) = \vartheta[F(X_0, \ldots, X_{n-1})]$, where $X'_k = \vartheta_{\sigma_k}(X_k)$ for $k < n$. This is well-defined, for if we also have $X'_k = \vartheta_{\sigma_k}(Y_k)$ for $k < n$, then using condition (1) for σ_k we see that $1 = \| X'_k \equiv X'_k \| = \| \vartheta_{\sigma_k}(X_k) \equiv \vartheta_{\sigma_k}(Y_k) \| = \vartheta \| X_k \equiv Y_k \|$. Since $F \in M_\sigma$, however, the inequality

(2) $\Pi_{k < n} N \| X_k \equiv Y_k \| \leq [F(X) \Leftrightarrow F(Y)]$

holds in B, and applying ϑ to this we have $1 = \Pi_{k < n} N' \vartheta \| X_k \equiv Y_k \| = \vartheta[\Pi_{k < n} N \| X_k \equiv Y_k \|] \leq \vartheta[F(X) \Leftrightarrow F(Y)] = [\vartheta[F(X)] \Leftrightarrow \vartheta[F(Y)]]$, so

---

[6] Such a mapping ϑ will preserve the subalgebra of open elements. However, a complete Boolean epimorphism which preserves the set of open elements may fail to preserve the operator N.

that $\vartheta[F(X)] = \vartheta[F(Y)]$. In essentially the same way we show that $\vartheta_\sigma(F)$ belongs to M'_σ. For suppose $X', Y' \in M'_{\sigma_0} \times \ldots \times M'_{\sigma_{n-1}}$, and let $X'_k = \vartheta_{\sigma_k}(X_k)$, $Y'_k = \vartheta_{\sigma_k}(Y_k)$ for $k < n$. Applying ϑ to inequality (2) again, and using condition (1) for σ_k, we have

$$\Pi_{k<n} \, N' \, \| X'_k \equiv Y'_k \| \leq [\vartheta[F(X)] \Leftrightarrow \vartheta[F(Y)]]$$
$$= [\vartheta_\sigma(F)(X') \Leftrightarrow \vartheta_\sigma(F)(Y')],$$

as required. Finally, we show that ϑ_σ is onto M'_σ: Suppose $F' \in M'_\sigma$, and define a mapping F from $M_{\sigma_0} \times \ldots \times M_{\sigma_{n-1}}$ into B by:

(3) $\quad F(X) = \Sigma_{P \in B, \, \vartheta(P) \leq F'(X')} \, P$,

where $X' = (X'_0, \ldots, X'_{n-1})$ and $X'_k = \vartheta_{\sigma_k}(X_k)$ ($k < n$). Applying ϑ to (3):

$$\vartheta[F(X)] = \Sigma_{P \in B, \, \vartheta(P) \leq F'(X')} \, \vartheta(P) = F'(X'),$$

since ϑ is onto B', so it remains only to verify that $F \in M_\sigma$, i.e., that (2) holds for arbitrary $X, Y \in M_{\sigma_0} \times \ldots \times M_{\sigma_{n-1}}$. For this it suffices to show that

$$\Pi_{k<n} \, N \, \| X_k \equiv Y_k \| \cdot F(X) \leq F(Y),$$

and by the definition of $F(Y)$ it suffices in turn to show

(4) $\quad \vartheta[\Pi_{k<n} \, N \, \| X_k \equiv Y_k \| \cdot F(X)] \leq F'(Y')$,

where $Y'_k = \vartheta_{\sigma_k}(Y_k)$ ($k < n$). But (4) is equivalent to

$$\Pi_{k<n} \, N' \, \| X'_k \equiv Y'_k \| \cdot F'(X') \leq F'(Y'),$$

where $X'_k = \vartheta_{\sigma_k}(X_k)$ ($k < n$), and this clearly holds since $F' \in M'_\sigma$.

To verify condition (1) for σ, suppose $F, G \in M_\sigma$ and let $\times M_{\sigma_k}$ denote the product $M_{\sigma_0} \times \ldots \times M_{\sigma_{n-1}}$. Then:

$$\vartheta \| F \equiv G \| = \vartheta[\Pi_{X \in \times M_{\sigma_k}} [F(X) \Leftrightarrow G(X)]]$$
$$= \Pi_{X \in \times M_{\sigma_k}} [\vartheta[F(X)] \Leftrightarrow \vartheta[G(X)]]$$

$$= \prod_{X' \in \times M'_{\sigma_k}} [\vartheta_\sigma(F)(X') \Leftrightarrow \vartheta_\sigma(G)(X')]$$

$$= \|\vartheta_\sigma(F) \equiv \vartheta_\sigma(G)\|,$$

as required. To complete the definition of the image model M', we define $m'(c_\sigma) = \vartheta_\sigma[m(c_\sigma)]$ for each constant c_σ. We can now state:

LEMMA 16.3. Let M be a (B,N)-valued topological model of ML_p, and let M' be the image of M under a complete homomorphism ϑ from (B,N) onto (B',N'). Then for every formula A of ML_p, every a ∈ As(M) and a' ∈ As(M'), if $a'(x_\sigma) = \vartheta_\sigma[a(x_\sigma)]$ for every variable x_σ then we have $\|A\|_{a'}^{M'} = \vartheta\|A\|_a^M$. In particular, if A is a closed formula of ML_p then $\|A\|^{M'} = \vartheta\|A\|^M$.

Proof: Straightforward, by induction on A.

Before applying Lemma 16.3 we need some algebraic preliminaries. The <u>direct product</u> (B,N) of a family of complete S5-algebras (B_i, N_i) (i ∈ I) is defined in the usual way: B is the ordinary direct product of the Boolean algebras B_i, $B = \times_{i \in I} B_i$, consisting of all indexed families $(P_i)_{i \in I}$ such that $P_i \in B_i$ for all i ∈ I, with the Boolean operations defined coordinate-wise,[7] and the interior operator N is defined by letting $N(P_i)_{i \in I} = (N_i P_i)_{i \in I}$. It is a routine matter to verify that (B,N) is again a complete S5-algebra, and that the <u>projection</u> mapping π_j from B onto B_j defined by $\pi_j(P_i)_{i \in I} = P_j$ is a complete homomorphism from (B,N) onto (B_j, N_j) for each index j ∈ I. Let us call a complete S5-algebra <u>reducible</u> if it is isomorphic to a direct product of trivial complete S5-algebras, i.e., algebras (B_i, N_i) in which the interior operator N_i is the trivial one: $N_i P = 1$ if $P = 1$, $N_i P = 0$ otherwise. We say that a (B,N)-valued topological model is <u>reducible</u> when the algebra (B,N) is reducible. It is not difficult to show that (B,N) is reducible if and only if the complete subalgebra B^* of open elements of (B,N) is atomic. We omit the proof.

THEOREM 16.4. A formula A of ML_p is b-valid if and only if it is true in every reducible topological model of ML_p.

---

[7] Sikorski calls this the <u>direct union</u>; see Sikorski [1969], p. 50.

Proof: Since every Boolean model of ML_p is a reducible (in fact, trivial) topological model, one implication is immediate. For the converse, assume A is b-valid. We can assume without loss of generality that A is closed, since A can be replaced by its universal closure otherwise. Let M be any (B,N)-valued topological model of ML_p, where (B,N) is a reducible complete S5-algebra. We can assume that (B,N) is itself the direct product of trivial S5-algebras (B_i, N_i) $(i \in I)$. If A is not true in M then $\| A \|^M \neq 1$, so clearly $\pi_i \| A \|^M \neq 1$ in B_i for some $i \in I$. Let M_i be the image of the model M under the complete homomorphism π_i. M_i is a (B_i, N_i)-valued topological model of ML_p, hence a Boolean model, since N_i is trivial. In view of the assumption that A is b-valid, we conclude that $\| A \| = 1$ in M_i. On the other hand, Lemma 16.3 implies that

$$\| A \|^{M_i} = \pi_i \| A \|^M \neq 1 ,$$

which is a contradiction.

Let (B,N) be any topological algebra, $P \in B$ a non-zero open element. The set $B_P = \{ Q \in B \mid Q \leq P \}$ becomes a Boolean algebra

$$(B_P, +_P, \cdot_P, -_P, 0_P, 1_P)$$

if we let $+_P$ and $\cdot_P$ coincide with the corresponding operations $+$, $\cdot$ of B, define $-_P Q$ to be $P \cdot (-Q)$, and let 0_P, 1_P be 0, P, respectively. If we let N_P be the operator on B_P which coincides with the operator N for all $Q \in B_P$, then (B_P, N_P) is a topological algebra, which is complete (resp., an S5-algebra) if (B,N) is complete (resp., an S5-algebra). The mapping $\vartheta(Q) = P \cdot Q$ is a complete homomorphism from (B,N) onto (B_P, N_P). Conversely, if (B,N) is a complete topological algebra and ϑ is a complete homomorphism from (B,N) onto (B',N'), then it is easily shown that (B',N') is isomorphic to (B_P, N_P), where P is the open element $\Pi_{Q \in B, \vartheta(Q) = 1} Q$ of B. It follows, in particular, that a trivial complete S5-algebra has no proper complete homomorphic images.

THEOREM 16.5. Let A be a modally closed formula of ML_p which is not t-valid. Then for some topological model M of ML_p and some $a \in As(M)$, we have $\| A \|^M_a = 0$.

Proof: Let M be a (B,N)-valued topological model, a an assignment over M such that $\| A \|_a^M \neq 1$. Since A is modally closed, the formula $\square A \leftrightarrow A$ is provable in ML_p and therefore true in M, by Theorem 16.2, so that the element $P = -\| A \|_a^M \neq 0$ is open in (B,N). Let ϑ be the complete homomorphism from (B,N) onto (B_p, N_p) defined earlier, and let M' be the image of M under ϑ, a' the assignment over M' defined by $a'(x_\sigma) = \vartheta_\sigma[a(x_\sigma)]$. Then by Lemma 16.3,

$$\| A \|_{a'}^{M'} = \vartheta \| A \|_a^M = P \cdot \| A \|_a^M = 0 ,$$

as required, and the proof is complete.

The hypothesis that A be modally closed cannot be dropped from Theorem 16.5. For example, the formula $[p_\phi \vee \square \sim p_\phi]$ fails to satisfy the conclusion of the theorem, but it is not even valid in ML_p, and therefore a fortiori not t-valid.

We conclude with several remarks. It can be shown that there exist b-valid formulas of ML_p which are not t-valid. For it follows from the results of Scott [1966] that there are sentences of ordinary higher-order logic which are valid (true in every standard model of L_p) but not true in every Boolean model, in his sense. By relativizing higher-order quantifiers to hereditarily constant predicates, we obtain a closed and modally closed formula A of ML_p which is valid in ML_p but fails in some discrete topological model M, in the sense that $\| A \| \neq 1$ in M. Since A is valid in ML_p, Theorem 15.10 implies that the formula $[At \to A]$ is b-valid; but if this formula were t-valid then it would be true in M, which is impossible since it is easily checked that $\| At \| = 1$ in any discrete complete S5-algebra.

The problem of classifying the various types of complete S5-algebras -- or equivalently, the various types of pairs (B, B^*) where B^* is a complete subalgebra of B -- is probably not trivial, in view of a result of Kripke[8] which implies that any complete Boolean algebra B^* is a complete subalgebra of some countably generated complete Boolean algebra B.

---

[8] Kripke [1967].

§17. Cohen's Independence Results

We have seen in §15 that the Boolean semantics for ML_p yields a number of independence results in modal logic; i.e., theorems to the effect that a particular modal formula is not provable in a particular modal theory. We can also use the Boolean semantics for ML_p to study certain of the well-known independence results in set theory first proved by Cohen.[1]

It was a joint discovery of Scott and Solovay[2] that Cohen's independence proofs could be recast in the framework of Boolean-valued logic, by means of Boolean models similar to those introduced in §14. In particular, one can show in this way that the continuum hypothesis cannot be proved in set theory even when the axiom of choice is added. In the first published exposition of the Boolean method,[3] Scott actually proved a somewhat weaker result, viz., that the continuum hypothesis cannot be proved in higher-order number theory with the axiom of choice. This proof involves all the ideas needed for the stronger result and avoids some of the messier details connected with Boolean models of the full theory of sets. We examine now a proof, similar to Scott's, of the independence of the continuum hypothesis in higher-order number theory with the axiom of choice. Our proof, however, employs formulas of ML_p as modal "interpolants" at various stages of the argument. It should be remarked that the present approach could be adapted to the context of set theory by replacing the higher-order modal logic ML_p by a full modal set theory.[4]

<u>Higher-Order Number Theory</u>. We augment the axioms of the theory L_p+C defined on pages 70-71 by several additional axioms to obtain a theory N_p, <u>Higher-Order Number Theory</u>, which will play the role of set theory in our formulation of the independence proof. We first add, for each type σ, an

---

[1] Cohen [1963].

[2] See Scott [1966], [1967a], [1967b], Rosser [1969].

[3] Scott [1967a].

[4] See Lemmon [1963].

axiom of choice:

Ac^σ : $\exists f_{(\sigma,\sigma)}\ Wo^\sigma(f)$,

where $Wo^\sigma(f)$ is the formula

$$\forall x_\sigma\ \forall y_\sigma\ [\ fxy \vee x \equiv y \vee fyx\ ] \wedge \forall g_{(\sigma)}\ [\ \exists x_\sigma\ gx \rightarrow$$
$$\exists x_\sigma\ [\ gx \wedge \forall y_\sigma\ [\ gy \rightarrow\ \sim fyx\ ]\ ]\ ] .$$

Ac^σ asserts that a well-ordering of the objects of type σ exists; it is therefore valid, i.e., true in all standard models of L_p.

Consider now a standard model of L_p based on the set ω of natural numbers. If we let $<$ be the first constant of type (e,e) and write $x < y$ for $<xy$, we can characterize the usual ordering on ω by the following second-order Peano axiom:

Pe : $Wo^e(<) \wedge \forall x_e\ \exists y_e\ [x < y] \wedge \forall y_e\ [\ \exists x_e\ [x < y] \rightarrow$
$$\exists x_e\ \forall z_e\ [\ z < y \leftrightarrow z < x \vee z \equiv x\ ]\ ] ,$$

which asserts that the usual ordering of ω is a well-ordering with no greatest element in which every element except the least has an immediate predecessor.

The theory N_p takes for its axioms all those axioms of L_p+C which contain no constants other than $<$, together with the Peano axiom Pe and the axioms Ac^σ for $\sigma \in P$. The rules of inference are those of the theory L_p (Rules R1, R2 on page 71). Clearly, any standard model of N_p is isomorphic to the standard model $M = (M_\sigma, m)_{\sigma \in P}$ based on ω , in which $m(<)$ is the usual ordering on ω .[5] We can easily define the usual number-theoretic operations as relations in N_p, using higher-order quantifiers, and rational and real numbers can be defined as higher-order entities and their usual properties established in N_p. Thus, the theory N_p incorporates the usual body of algebra and analysis normally required in mathematical arguments.

The continuum hypothesis is the assertion that every class of subsets of ω is either denumerable or else equinumerous with $P(\omega)$. Since the

---

[5] Here we ignore the values $m(c_\sigma)$ for constants c_σ other than $<$.

axiom of choice is present, we can formulate this as follows: Every non-empty subset G of $P(\omega)$ is either the range of a function on ω or else the domain of a function onto $P(\omega)$. We therefore seek to establish that the formula Ch below is not provable in the theory N_p:

Ch : $\forall g_{((e))} [\exists y_{(e)} \; gy \rightarrow :$

$\exists f_{(e,(e))} [\forall x_e \; \exists!y_{(e)} \; [gy \wedge fxy] \wedge \forall y_{(e)} [gy \rightarrow \exists x_e \; fxy]]$

$\vee \; \exists f_{((e),(e))} [\forall y_{(e)} [gy \rightarrow \exists!z_{(e)} \; fyz] \wedge \forall z_{(e)} \exists y_{(e)} [gy \wedge fyz]]]$.

<u>Extensional Predicates</u>. We introduced at the end of §1 a distinction between extensional and intensional words of English, and we indicated how the extensionality, for example, of an adjective ζ could be expressed by a semantic condition on its intension Int$[\zeta]$ in a model of formalized English. This distinction can be carried over to the formal logic ML_p, where it proves to be an important element in the independence proof.

Let $M = (M_\sigma, m)_{\sigma \in P}$ be a g-model of ML_p. For each type $\sigma \neq e$ we single out a class E_σ of <u>(hereditarily) extensional</u> predicates $F \in M_\sigma$. For convenience, we define E_σ also for $\sigma = e$, so that the definition takes the following form:

(i) $\quad E_e = D = M_e$,

(ii) $\quad$ For $\sigma = (\sigma_0, \ldots, \sigma_{n-1})$, E_σ consists of all $F \in M_\sigma$ such that
(a) $F(i) \subseteq E_{\sigma_0} \times \ldots \times E_{\sigma_{n-1}}$ for all $i \in I$; (b) For any $X, Y \in E_{\sigma_0} \times \ldots \times E_{\sigma_{n-1}}$ and any $i \in I$, if $X_k(i) = Y_k(i)$ when $\sigma_k \neq e$ and $X_k = Y_k$ when $\sigma_k = e$, then $X \in F(i)$ iff $Y \in F(i)$.

That is, the question whether a given sequence $(X_0, \ldots, X_{n-1})$ of extensional objects belongs to $F(i)$ depends only on the values $X_k(i)$ where $\sigma_k \neq e$ and the values X_k where $\sigma_k = e$. We can define the classes E_σ formally in ML_p. Precisely, we define for each $\sigma \in P$ and each variable x_σ a formula

$\quad$ Ext$^\sigma(x)$ $\quad$ (x <u>is extensional</u>)

of ML_p with exactly x_σ free, as follows:

(i) $\quad$ Ext$^e(x_e)$ is $[x \equiv x]$,

(ii) For $\sigma = (\sigma_0,\ldots,\sigma_{n-1})$, $\text{Ext}^\sigma(f_\sigma)$ is the formula

$$\Box \, \forall x^0 \ldots \forall x^{n-1} \, [\, f \, x^0 \ldots x^{n-1} \to \text{Cnj}_{k<n} \, \text{Ext}^{\sigma_k}(x^k) \,] \, \wedge$$

$$\Box \, \forall x^0 \ldots \forall x^{n-1} \, \forall y^0 \ldots \forall y^{n-1} \, [\, \text{Cnj}_{k<n} \, [\, \text{Ext}^{\sigma_k}(x^k) \wedge$$

$$\text{Ext}^{\sigma_k}(y^k) \wedge x^k \equiv y^k \,] \to [\, f \, x^0 \ldots x^{n-1} \leftrightarrow f \, y^0 \ldots y^{n-1} \,] \,] \, ,$$

where $x^0, \ldots, x^{n-1}, y^0, \ldots, y^{n-1}$ is the first sequence of distinct variables such that x^k and y^k are of type σ_k $(k < n)$.[6]

The formula $\text{Ext}^\sigma(x)$ is modally closed, in the extended sense that the equivalence $\Box \, \text{Ext}^\sigma(x) \leftrightarrow \text{Ext}^\sigma(x)$ is provable in ML_p, and if M is a g-model of ML_p and $X \in M_\sigma$ then clearly $X \in E_\sigma$ if and only if M satisfies $\text{Ext}^\sigma(X)$. We shall henceforth drop the superscript on the formula $\text{Ext}^\sigma(x)$, the type σ being clear from context.

LEMMA 17.1. The formulas $\forall x_\sigma \, \text{Ext}(x)$ are provable in ML_p for $\sigma = e$ and $\sigma = \bar{n}$ $(n \in \omega)$, and the formula $\exists x_\sigma \, \text{Ext}(x)$ is provable in $ML_p + C$ for every type σ.

The proof is omitted.

If $M = (M_\sigma, m)_{\sigma \in P}$ is a B-valued Boolean model of ML_p based on D, where B is a complete Boolean algebra, then for each $X \in M_\sigma$ we clearly have $\| \text{Ext}(X) \| = 0$ or 1, and if we let $E_\sigma = \{ X \in M_\sigma \mid \| \text{Ext}(X) \| = 1 \}$, we obtain the following recursive characterization:

(i) $E_e = D = M_e$,

(ii) For $\sigma = (\sigma_0,\ldots,\sigma_{n-1})$, E_σ consists of all $F \in M_\sigma$ such that
 (a) For $X \in M_{\sigma_0} \times \ldots \times M_{\sigma_{n-1}}$, $F(X) = 0$ unless $X_k \in E_{\sigma_k}$ $(k < n)$;
 (b) For $X, Y \in E_{\sigma_0} \times \ldots \times E_{\sigma_{n-1}}$, F satisfies the inequality
 $\Pi_{k < n} \, \| X_k \equiv Y_k \| \leq [F(X) \Leftrightarrow F(Y)]$.

Thus, if $m(c_\sigma) \in E_\sigma$ for each constant c_σ, then the extensional objects in M form a B-valued general Boolean model of ML_p, in the sense of §16,

---

[6] Here $\text{Cnj}_{k<n} \, A^k$ denotes the conjunction $[A^0 \wedge \ldots \wedge A^{n-1}]$.

which is isomorphic to a (B,N)-valued topological model, where N is the discrete operator on B.

Given a formula A of L_p, we let $A^{(Ext)}$ denote the formula of ML_p obtained by relativizing all quantifiers of A to the formula $Ext(x)$; i.e., replacing each subformula $\forall x_\sigma \ B$ of A by $\forall x_\sigma \ [Ext(x) \to B]$. Let Δ_A consist of all formulas $Ext(s_\sigma)$, where s_σ is either a constant in A or a free variable of A . The next theorem shows that, loosely speaking, the extensional predicates in a model of higher-order modal logic form a model of ordinary (non-modal) higher-order logic. Precisely:

THEOREM 17.2. If A is provable in L_p+C then $A^{(Ext)}$ is derivable from Δ_A in ML_p+C .

Proof: One shows that the class of formulas A for which $A^{(Ext)}$ is derivable from Δ_A contains all the axioms of L_p+C and is closed under the rules R1, R2 of L_p. As in the proof of Theorem 16.2, it is necessary to use the fact that axiom AS5 of L_p (page 71) can be replaced by:

A5.1 $x_e \equiv y_e \to y_e \equiv x_e$,

A5.2 $x_e \equiv y_e \to [y_e \equiv z_e \to x_e \equiv z_e]$,

A5.3 $x^0 \equiv y^0 \wedge \ldots \wedge x^{n-1} \equiv y^{n-1} \to [f \ x^0 \ldots x^{n-1} \to f \ y^0 \ldots y^{n-1}]$,
where $x^0, \ldots, x^{n-1}, y^0, \ldots, y^{n-1}$ is the first sequence of distinct variables such that x^k and y^k are of type σ_k $(k < n)$, and f is of type $\sigma = (\sigma_0, \ldots, \sigma_{n-1})$.

In addition, the proof makes use of Lemma 17.1, and (for the verification of A5.3) the following result, which we state without proof:

LEMMA 17.2.1. For any type σ the formula

$$Ext(x_\sigma) \wedge Ext(y_\sigma) \to [[x \equiv y]^{(Ext)} \leftrightarrow x \equiv y]$$

is provable in ML_p.

Modal Number Theory. We consider now a modal analogue of N_p, which we denote by MN_p. Its axioms are those axioms of ML_p+C containing only the constant < of type (e,e) , together with the <u>axioms of choice</u>:

AC^σ : $\exists f_{(\sigma,\sigma)}\ WO^\sigma(f)$,

where $WO^\sigma(f)$ is the formula

$$Rn(f) \wedge \forall x_\sigma \forall y_\sigma [\ fxy \vee x \equiv y \vee fyx\ ] \wedge$$
$$\Box \forall g_{(\sigma)} [\ \exists x_\sigma\ gx \rightarrow \exists x_\sigma [\ gx \wedge \forall y_\sigma [\ gy \rightarrow\ \sim fyx\ ]\ ]\ ] ,$$

and the Peano axiom

PE : $WO^e(<) \wedge \forall x_e \exists y_e\ [x < y] \wedge \forall y_e [\ \exists x_e\ [x < y]\ \rightarrow$
$$\exists x_e \forall z_e [\ z < y \leftrightarrow z < x \vee z \equiv x\ ]\ ] .$$

The inference rules of MN_p are the rules R1, R2, R3 of ML_p (page 74). It is clear that the axioms of MN_p will all hold true in the standard model $M = (M_\sigma, m)_{\sigma \in P}$ of ML_p based on ω and any set I of indices, if we take for $m(<)$ the function on I whose value at any $i \in I$ is the usual ordering on ω. To see that AC^σ holds, it suffices to take any well-ordering $(X_\xi)_{\xi < \lambda}$ of M_σ and define $F \in M_{(\sigma,\sigma)} = P(M_\sigma \times M_\sigma)^I$ by putting $F(i) = \{\ (X_\xi, X_\eta)\ |\ \xi < \eta\ \}$ $(i \in I)$.

LEMMA 17.3.1. The formula $AC^\sigma \rightarrow [Ac^\sigma]^{(Ext)}$ is provable in $ML_p + C$ for each type $\sigma \in P$.

Proof: Let $M = (M_\sigma, m)_{\sigma \in P}$ be a g-model of $ML_p + C$ based on D and I such that M sat AC^σ,[7] and suppose $i \in I$. Say $F \in M_{(\sigma,\sigma)}$ and M sat $WO^\sigma(F)$. We show that M, i sat $[Ac^\sigma]^{(Ext)}$, or equivalently, that

$$M,\ i\ \text{sat}\ \exists f_{(\sigma,\sigma)} [\ Ext(f) \wedge Wo^\sigma(f)^{(Ext)}\ ] .$$

By comprehension there exists $F' \in M_{(\sigma,\sigma)}$ such that M satisfies

(1) $\Box \forall x_\sigma \forall y_\sigma [\ F'xy \leftrightarrow Ext(x) \wedge Ext(y) \wedge \forall y'_\sigma [\ Ext(y') \wedge$
$$y \equiv y' \rightarrow \exists x'_\sigma [\ Ext(x') \wedge x \equiv x' \wedge Fx'y'\ ]\ ]\ ] .$$

It is easy to see from (1) that M sat $Ext(F')$, so it remains only to show that M, i sat $Wo^\sigma(F')^{(Ext)}$. By Lemma 17.2.1 and some obvious sim-

---

[7] We regard the formulas AC^σ and PE as modally closed, in view of the fact that $[\Box AC^\sigma \leftrightarrow AC^\sigma]$ and $[\Box PE \leftrightarrow PE]$ are theorems of ML_p.

plifications, it suffices in turn to show that M, i satisfy

(2) $\forall x_\sigma \forall y_\sigma$ [Ext(x) ∧ Ext(y) → F'xy ∨ x ≡ y ∨ F'yx] ∧

$\forall g_{(\sigma)}$ [Ext(g) ∧ $\exists x_\sigma$ gx → $\exists x_\sigma$ [gx ∧ $\forall y_\sigma$ [gy → ~ F'yx]]] .

For the first conjunct, assume X , Y ∈ E_σ and suppose that M, i do not satisfy X ≡ Y or F'YX . We show using (1) that M, i sat F'XY . Since M, i do not satisfy F'YX , there exists X' ∈ E_σ such that M, i sat X ≡ X' and for every Y' ∈ E_σ , if M, i sat Y ≡ Y' then (Y',X') is not in F(i) . Hence if Y' ∈ E_σ and M, i sat Y ≡ Y' then (X',Y') ∈ F(i) , because M sat $WO^\sigma(F)$ and M, i do not satisfy X ≡ Y , therefore do not satisfy X' ≡ Y' or, a fortiori, X' ≡ Y' . This shows that M, i sat F'XY . For the second conjunct of (2), assume G ∈ $E_{(\sigma)}$ and G(i) ≠ ϕ . Since M sat $WO^\sigma(F)$, there exists X ∈ M_σ , X ∈ G(i) , such that for all Y ∈ M_σ , Y ∈ G(i) implies (Y,X) ∉ F(i) . But then, using extensionality and (1), Y ∈ G(i) implies (Y,X) ∉ F'(i) . This completes the proof.

LEMMA 17.3.2. The formula PE → $Pe^{(Ext)}$ is provable in ML_p+C .

Proof: Using Lemma 17.1 it is easily shown that $Pe^{(Ext)}$ is equivalent in ML_p to Pe , and clearly PE implies Pe in ML_p.

THEOREM 17.3. If a closed formula A is a theorem of N_p, then $A^{(Ext)}$ is a theorem of MN_p.

Proof: Suppose A is a theorem of N_p. Then for some types σ_0 , ... , σ_{n-1} the formula

$$Ac^{\sigma_0} \to . Ac^{\sigma_1} \to . \ldots \to . Ac^{\sigma_{n-1}} \to . Pe \to A$$

is provable in L_p+C , without using constants other than < . By Theorem 17.2, therefore, the formula

$$[Ac^{\sigma_0}]^{(Ext)} \to . \ldots \to . [Ac^{\sigma_{n-1}}]^{(Ext)} \to . Pe^{(Ext)} \to A^{(Ext)}$$

is derivable (in fact, without using constants other than <) from the formula Ext(<) in ML_p+C . But by Lemma 17.1, Ext(<) is a theorem of ML_p. The result now follows by Lemmas 17.3.1 and 17.3.2.

We now introduce two closed and modally closed formulas of ML_p which will act as the modal interpolants mentioned earlier. The first is the formula

$$C_1 : \quad \forall f_{(e,(e))} \; [\; \Box \, \forall x_e \, \forall y_{(e)} \, \forall y'_{(e)} \; [\; Rn(y) \wedge Rn(y') \wedge fxy \wedge$$
$$fxy' \to y \equiv y' \;] \; \to \; \exists y_{(e)} \; [\; Rn(y) \wedge \Box \, \forall x_e \sim fxy \;] \;] \;,$$

which can be given the following reading: If a predicate of type $(e,(e))$ is necessarily functional from individuals to sets (of individuals), then some set is necessarily out of its range. The second interpolant is the formula

$$C_2 : \quad \forall f_{((e),(e))} \; [\; \Box \, \forall y_{(e)} \, \forall z_{(e)} \, \forall z'_{(e)} \; [\; Rn(y) \wedge fyz \wedge$$
$$fyz' \to z \equiv z' \;] \; \to \; \exists z_{(e)} \, \Box \, \forall y_{(e)} \; [\; Rn(y) \to \sim fyz \;] \;] \;,$$

which can be rendered (somewhat less satisfactorily) as follows: If a predicate of type $((e),(e))$ is necessarily functional from sets to properties then some property is necessarily out of its range.

It is not difficult to show that for a standard model M of ML_p based on sets D and I, M sat C_1 just in case $|I| \cdot |D| < 2^{|D|}$, so that in particular if D is infinite then M sat C_1 just in case $|I| < 2^{|D|}$. On the other hand, a standard model never satisfies C_2; i.e., the formula $\sim C_2$ is valid in ML_p.

LEMMA 17.4. *The formula* $\sim Ch^{(Ext)}$ *is derivable from* C_1 *and* C_2 *in* $ML_p + C$.

Proof: $Ch^{(Ext)}$ is equivalent in ML_p to the following formula:[8]

$$\forall g_{((e))} \; [\; Ext(g) \wedge \exists y_{(e)} \, gy \; \to \; :$$
$$\exists f_{(e,(e))} \; [\; Ext(f) \wedge \forall x_e \, \exists ! y_{(e)} \; [gy \wedge fxy]$$
$$\wedge \forall y_{(e)} \; [\; gy \to \exists x_e \, fxy \;] \;]$$
$$\vee \; \exists f_{((e),(e))} \; [\; Ext(f) \wedge \forall y_{(e)} \; [\; gy \to \exists ! z_{(e)} \, fyz \;]$$
$$\wedge \forall z_{(e)} \, \exists y_{(e)} \, [gy \wedge fyz] \;] \;] \;.$$

---

[8] Note the meaning of $\exists ! x \, A$ here; see pp. 69, 71 in §9.

Suppose M is a g-model of ML_p+C such that M sat C_1, M sat C_2, and for some $i \in I$, M, i sat $Ch^{(Ext)}$. By comprehension in M, there exists $G \in M_{((e))}$ such that M satisfies

$$\Box \forall y_{(e)} [Gy \leftrightarrow \exists y'_{(e)} [Rn(y') \wedge y \equiv y']].$$

The following properties of G are easily verified:

(1) M sat $\Box \forall y_{(e)} [Rn(y) \to Gy]$,

(2) M sat $Ext(G)$; i.e., $G \in E_{((e))}$,

(3) M, i sat $\exists y_{(e)} Gy$ (using (1)).

But M, i sat $Ch^{(Ext)}$; we therefore have two cases:

Case 1. For some $F \in E_{(e,(e))}$,

(4) M, i sat $\forall x_e \exists! y_{(e)} [Gy \wedge Fxy]$

(5) M, i sat $\forall y_{(e)} [Gy \to \exists x_e Fxy]$

Case 2. For some $F \in E_{((e),(e))}$,

(6) M, i sat $\forall y_{(e)} [Gy \to \exists! z_{(e)} Fyz]$

(7) M, i sat $\forall z_{(e)} \exists y_{(e)} [Gy \wedge Fyz]$.

First suppose Case 1 holds. By comprehension, there exists $F' \in M_{(e,(e))}$ such that

(8) M sat $\Box \forall x_e \forall y_{(e)} [F'xy \leftrightarrow Fxy \wedge \forall x_e \exists! y_{(e)} [Gy \wedge Fxy]]$.

We claim that M satisfies the formula

$$\Box \forall x_e \forall y_{(e)} \forall y'_{(e)} [Rn(y) \wedge Rn(y') \wedge F'xy \wedge F'xy' \to y \equiv y'].$$

For, let $j \in I$, $X \in M_e$, $Y, Y' \in M_{(e)}$ such that

M, j sat $Rn(Y) \wedge Rn(Y') \wedge F'XY \wedge F'XY'$.

Then by (1), M, j satisfy GY and GY', and by (8), M, j satisfy $\forall x_e \exists! y_{(e)} [Gy \wedge Fxy]$ and also M, j sat FXY, FXY', so M, j sat $Y \equiv Y'$. This proves the claim. Since M sat C_1, we conclude that there exists $Y \in M_{(e)}$ such that M sat $[Rn(Y) \wedge \Box \forall x_e \sim F'xY]$. But by (1),

M, i sat GY , so by (5), M, i sat $\exists x_e$ FxY , whence by (4) and (8) we conclude that M, i sat $\exists x_e$ F'xY , a contradiction. Suppose, on the other hand, that Case 2 holds. By comprehension, there exists F' $\in$ $M_{((e),(e))}$ such that

(9)   M sat $\Box \forall y_{(e)} \forall z_{(e)}$ [F'yz $\leftrightarrow$ Fyz $\wedge \forall y_{(e)}$ [Gy $\to \exists ! z_{(e)}$ Fyz]] .

We now claim that M satisfies the formula

$$\Box \forall y_{(e)} \forall z_{(e)} \forall z'_{(e)} [Rn(y) \wedge F'yz \wedge F'yz' \to z \equiv z'] .$$

For, let j $\in$ I , Y , Z , Z' $\in$ $M_{(e)}$ such that

M, j sat Rn(Y) $\wedge$ F'YZ $\wedge$ F'YZ' .

By (1), M, j sat GY , and by (9), M, j sat $\forall y_{(e)}$ [Gy $\to \exists ! z_{(e)}$ Fyz] , and also M, j sat FYZ , FYZ' , so M, j sat Z $\equiv$ Z' , which proves the claim. Since M sat C_2 , we conclude that there exists Z $\in$ $M_{(e)}$ such that

(10)   M sat $\Box \forall y_{(e)}$ [Rn(y) $\to \sim$ F'yZ] .

By (7), there exists Y $\in$ $M_{(e)}$ such that M, i sat GY $\wedge$ FYZ , and the choice of G implies that there exists Y' $\in$ $M_{(e)}$ such that M, i sat Rn(Y') $\wedge$ Y $\equiv$ Y' . Since F is extensional, M, i sat FY'Z , and hence by (6) and (9), M, i sat F'Y'Z , contradicting (10). This completes the proof of Lemma 17.4.

It should be remarked that Theorem 17.3 and Lemma 17.4 are syntactic results, for which direct, constructive proofs could be provided.

LEMMA 17.5.1. *The formula* AC^σ *is b-valid in* ML_p *for every type* σ .

Proof: Let M = $(M_\sigma, m)_{\sigma \in P}$ be a B-valued Boolean model of ML_p. Let $(X_\xi)_{\xi < \lambda}$ be a well-ordering of M_σ , and define F $\in$ $M_{(\sigma,\sigma)}$ by letting $F(X_\xi, X_\eta) = 1$ if $\xi < \eta$, and 0 otherwise. We show that $\| WO^\sigma(F) \| = 1$. Clearly the first two conjuncts of $WO^\sigma(F)$ have value 1 , so it suffices to assume G $\in$ $M_{(\sigma)} = B^{M_\sigma}$ and show that

(1) $\| \exists x_\sigma Gx \| \leq \| \exists x_\sigma [Gx \wedge \forall y_\sigma [Gy \to \sim Fyx]] \|$.

But if we let $P_\xi = G(X_\xi)$ for $\xi < \lambda$, (1) is clearly equivalent to

(2) $\quad \Sigma_{\xi < \lambda} P_\xi \leq \Sigma_{\xi < \lambda} [P_\xi \cdot \Pi_{\eta < \xi} -P_\eta]$,

and (2) holds in any complete Boolean algebra.

LEMMA 17.5.2. Let $M = (M_\sigma, m)_{\sigma \in P}$ be a B-valued Boolean model of ML_P based on ω, and let $m(<) \in M_{(e,e)} = B^{\omega \times \omega}$ be the usual (two-valued) ordering on ω. Then $\|PE\|^M = 1$.

Proof: Straightforward.

THEOREM 17.5. The theory $MN_P + C_1 + C_2$ is consistent.

Proof: If not then some formula

(1) $\quad AC^{\sigma_0} \wedge AC^{\sigma_1} \wedge \ldots \wedge AC^{\sigma_{n-1}} \wedge PE \wedge C_1 \wedge C_2$

is inconsistent in $ML_P + C$. We construct a Boolean model of ML_P in which (1) has Boolean value 1, contradicting Theorem 14.3. Let I be any set with $|I| > 2^\omega$, and let B be the complete algebra of all regular open sets in the product space $2^{\omega \times I}$, where $2 = \{0,1\}$ has the discrete topology. It is well-known[9] that B satisfies the countable chain condition; i.e., every disjoint set of non-zero elements of B is denumerable. Let $M = (M_\sigma, m)_{\sigma \in P}$ be the B-valued Boolean model based on ω in which $m(<)$ is the usual ordering on ω. By Lemmas 17.5.1 and 17.5.2, we have $\|AC^{\sigma_k}\| = \|PE\| = 1$ in M, for all $k < n$, so it remains only to verify that $\|C_1\| = \|C_2\| = 1$.

If $\|C_1\| = 0$ (note that C_1 is modally closed), then there exists $F \in M_{(e,(e))}$ for which

(1) $\quad \|\Box \forall x_e \forall y_{(e)} \forall y'_{(e)} [Rn(y) \wedge Rn(y') \wedge Fxy \wedge Fxy' \to y \equiv y']\| = 1$,

(2) $\quad \|\exists y_{(e)} [Rn(y) \wedge \Box \forall x_e \sim Fxy]\| = 0$.

From (2) and Lemma 15.4.1, it follows that there exists for every $Y \in 2^\omega$ a natural number $n_Y \in \omega$ with $F(n_Y, Y) \neq 0$. For some $n \in \omega$, the set

---

[9] Rosser [1969], p. 32.

$K = \{ Y \in 2^\omega \mid n_Y = n \}$ must be non-denumerable. But by (1), $Y \neq Y'$ in K implies $F(n,Y) \cdot F(n,Y') = 0$, so that $\{ F(n,Y) \mid Y \in K \}$ is a non-denumerable disjoint set of non-zero elements of B, contradicting the countable chain condition.

If $\| C_2 \| = 0$ then for some $F \in M_{((e),(e))}$ we have:

(3) $\| \Box \forall y_{(e)} \forall z_{(e)} \forall z'_{(e)} [Rn(y) \wedge Fyz \wedge Fyz' \rightarrow z \equiv z'] \| = 1$,

(4) $\| \exists z_{(e)} \Box \forall y_{(e)} [Rn(y) \rightarrow \sim Fyz] \| = 0$.

For each $i \in I$ define $Z_i \in M_{(e)} = B^\omega$ by letting $Z_i(n)$ be the clopen subset $\{ p \in 2^{\omega \times I} \mid p(n,i) = 1 \}$ of the product space $2^{\omega \times I}$. It is easily verified that

$$\| Z_i \equiv Z_j \| = \prod_{n \in \omega} [Z_i(n) \Leftrightarrow Z_j(n)] = 0$$

for $i \neq j$ in I. By (4), there exists for each $i \in I$ some $Y_i \in 2^\omega$ with $F(Y_i, Z_i) \neq 0$. Since $|I| > 2^\omega$, there exists $Y \in 2^\omega$ such that the set $J = \{ i \in I \mid Y_i = Y \}$ is non-denumerable. By (3), $i \neq j$ in J implies

$$F(Y, Z_i) \cdot F(Y, Z_j) \leq \| Z_i \equiv Z_j \| = 0,$$

so that $\{ F(Y, Z_i) \mid i \in J \}$ is again a non-denumerable set of pairwise disjoint elements of B, contradicting the countable chain condition.

COROLLARY 17.6. The continuum hypothesis Ch is not provable in N_p.

Proof: If Ch were provable in N_p then $Ch^{(Ext)}$ would be provable in MN_p, by Theorem 17.3. It follows from Lemma 17.4 that the theory $MN_p + C_1 + C_2$ would then be inconsistent, contradicting Theorem 17.5.

BIBLIOGRAPHY

Peter Andrews

[1963] A reduction of the axioms for the theory of propositional types. Fund. Math., 52 (1963), 345-350.

Yehoshua Bar-Hillel

[1954] Indexical expressions. Mind, 63 (1954), 359-379.

Arnould Bayart

[1958] Correction de la logique modale du premier et du second ordre S5. Logique et analyse, 1 (1958), 28-44.

[1959] Quasi-adéquation de la logique modale du second ordre S5 et adéquation de la logique modale du premier ordre S5. Logique et analyse, 2 (1959), 99-121.

Aldo Bressan

[1964] A general many sorted modal language ML (Contributed paper, reported in Logic, methodology and philosophy of science: proceedings of the 1964 international congress (Amsterdam, 1965)).

[1972] A general interpreted modal calculus (New Haven, 1972).

Rudolf Carnap

[1947] Meaning and necessity (Chicago, 1947).

Alonzo Church

[1940] A formulation of the simple theory of types. J. Symb. Logic, 5 (1940), 56-68.

[1951] A formulation of the logic of sense and denotation. Structure, method and meaning, ed. Paul Henle et al. (New York, 1951).

[1956] Introduction to mathematical logic, vol. I (Princeton, 1956).

Nino Cocchiarella

[1966] Tense logic: a study of temporal reference (Doctoral dissertation, University of California at Los Angeles, 1966).

[1969] A completeness theorem in second order modal logic. Theoria, 2 (1969), 81-103.

Paul J. Cohen

[1963] The independence of the continuum hypothesis. Proc. Natl. Acad. Sci. U.S.A., 50 (1963), 1143-1148; 51 (1964), 105-110.

M. J. Cresswell

[1967] A Henkin completeness theorem for T. Notre Dame J. Form. Logic, 8 (1967), 186-190.

Donald Davidson and Gilbert Harman, eds.

[1972] Semantics of natural language (Dordrecht, 1972).

Herbert B. Enderton

[1972] A mathematical introduction to logic (New York and London, 1972).

Herbert Feigl and Wilfrid Sellars, eds.

[1949] Readings in philosophical analysis (New York, 1949).

Kit Fine

[1970] Propositional quantifiers in modal logic. Theoria, 36 (1970), 336-346.

Gottlob Frege

[1892] Über Sinn und Bedeutung. Zeitschrift für Philosophie und philosophische Kritik, 100 (1892), 25-50.

Paul R. Halmos

[1963] Lectures on Boolean algebras (Princeton, 1963).

Leon Henkin

[1950] Completeness in the theory of types. J. Symb. Logic, 15 (1950), 81-91.

[1963] A theory of propositional types. Fund. Math., 52 (1963), 323-344.

K. J. J. Hintikka

[1961] Modality and quantification. Theoria, 27 (1961), 119-128.

G. E. Hughes and M. J. Cresswell

[1968] An introduction to modal logic (London, 1968).

Stig Kanger

[1957] Provability in logic (Stockholm, 1957).

David Kaplan

[1964] Foundations of intensional logic (Doctoral dissertation, University of California at Los Angeles, 1964).

[1966] Review of Kripke [1963a]. J. Symb. Logic, 31 (1966), 120-122.

[1970] S5 with quantifiable propositional variables (Abstract), J. Symb. Logic, 35 (1970), 355.

Saul A. Kripke

[1959] A completeness theorem in modal logic. J. Symb. Logic, 24 (1959), 1-14.

[1963a] Semantical analysis of modal logic I. Z. Math. Logik Grundlagen Math., 9 (1963), 67-96.

[1963b] Semantical considerations on modal logic. Acta Philosophica Fennica, 16 (1963), 83-94.

[1967] An extension of a theorem of Gaifman-Hales-Solovay. Fund. Math. 61 (1967), 29-32.

E. J. Lemmon

[1963] A theory of attributes based on modal logic. Acta Philosophica Fennica, 16 (1963), 95-121.

David Lewis

[1970] General semantics. Synthese, 22 (1970), 18-67. Reprinted in Davidson and Harman [1972].

J. C. C. McKinsey

[1941] A solution of the decision problems for the Lewis systems S2 and S4, with an application to topology. J. Symb. Logic, 6 (1941), 117-134.

J. C. C. McKinsey and Alfred Tarski

[1944] The algebra of topology. Ann. of Math., 45 (1944), 141-191.

[1948] Some theorems about the sentential calculi of Lewis and Heyting. J. Symb. Logic, 13 (1948), 1-15.

Richard Montague

[1960] Logical necessity, physical necessity, ethics, and quantifiers. Inquiry, 4 (1960), 259-269. Reprinted in Montague [1974].

[1968] Pragmatics. Contemporary philosophy - la philosophie contemporaine, ed. Raymond Klibansky (Florence, 1968). Reprinted in Montague [1974].

[1969]   On the nature of certain philosophical entities. The Monist, 53 (1969), 159-194. Reprinted in Montague [1974].

[1970a] Pragmatics and intensional logic. Synthese, 22 (1970), 68-94. Reprinted in Davidson and Harman [1972] and Montague [1974].

[1970b] English as a formal language I. Linguaggi nella societa e nella tecnica, ed. Bruno Visentini et al. (Milan, 1970). Reprinted in Montague [1974].

[1970c] Universal grammar. Theoria, 36 (1970), 373-398. Reprinted in Montague [1974].

[1973]   The proper treatment of quantification in ordinary English. Approaches to natural language, ed. K. J. J. Hintikka et al. (Dordrecht, 1973), 221-242. Reprinted in Montague [1974].

[1974]   Formal philosophy: selected papers of Richard Montague, ed. Richmond Thomason (New Haven, 1974).

Andrzej Mostowski

[1947]   On absolute properties of relations. J. Symb. Logic, 12 (1947), 33-42.

Steven Orey

[1959]   Model theory for the higher order predicate calculus. Trans. Amer. Math. Soc., 92 (1959), 72-84.

Terence Parsons

[1968]   A semantics for English (Duplicated, 1968). An abridged version appears in Davidson and Harman [1972].

Barbara H. Partee

[1973]   The semantics of belief sentences. Approaches to Natural Language, ed. K. J. J. Hintikka et al. (Dordrecht, 1973), 309-336.

[1975a] Montague grammar and transformational grammar. To appear in Linguistic Inquiry, 6 (1975).

Barbara H. Partee, ed.

[1975b] Montague grammar (New York and London: Academic Press, to appear).

W. V. Quine

[1960]   Word and object (Cambridge, Mass., 1960).

Helena Rasiowa

[1951]   Algebraic treatment of the functional calculi of Lewis and Langford. Fund. Math., 38 (1951), 99-126.

[1963]   On modal theories. Acta Philosophica Fennica, 16 (1963), 201-214.

J. Barkley Rosser

[1969]   Simplified independence proofs: Boolean valued models of set theory (New York and London, 1969).

Dana Scott

[1966]   Boolean-valued models for higher-order logic (Duplicated notes, Stanford University, January 1966).

[1967a] A proof of the independence of the continuum hypothesis. Math. Systems Theory, 1 (1967), 89-111.

[1967b] Boolean-valued models for set theory (Lecture notes for the Amer. Math. Soc. summer institute on axiomatic set theory, University of California at Los Angeles, 1967).

[1970]   Advice on modal logic. Philosophical problems in logic: some recent developments, ed. Karel Lambert (Dordrecht, 1970), 143-173.

Roman Sikorski

[1969]   Boolean algebras, 3rd ed. (Berlin, 1969).

Alfred Tarski

[1954]   Contributions to the theory of models I. Indag. Math., 16 (1954), 572-581.

A. Tarski, A. Mostowski and R. M. Robinson

[1953]   Undecidable Theories (Amsterdam, 1953).

Richmond Thomason

[1970]   Some completeness results for modal predicate calculi. Philosophical problems in logic: some recent developments, ed. Karel Lambert (Dordrecht, 1970), 56-76.